DIY Microcontroller Projects for Hobbyists

The ultimate project-based guide to building real-world embedded applications in C and C++ programming

Miguel Angel Garcia-Ruiz

Pedro Cesar Santana Mancilla

BIRMINGHAM—MUMBAI

DIY Microcontroller Projects for Hobbyists

Copyright © 2021 Packt Publishing

Group Product Manager: Richa Tripathi

Publishing Product Manager: Shweta Bairoliya

Senior Editor: Storm Mann

Content Development Editor: Kinnari Chohan

Technical Editor: Karan Solanki

Copy Editor: Safis Editing

Project Coordinator: Deeksha Thakkar

Proofreader: Safis Editing

Indexer: Tejal Daruwale Soni

Production Designer: Nilesh Mohite

First published: June 2021

Production reference: 1290621

Published by Packt Publishing Ltd.

Livery Place

35 Livery Street

Birmingham

B3 2PB, UK.

ISBN 978-1-80056-413-8

www.packt.com

To my parents, Rosa and Miguel, for giving me the love to learn new things.
To my wife, Selene, and my son, Miguel, for their support, love,
and inspiration.

– Miguel Garcia-Ruiz

To my son, Max, for his love and inspiration, and for making my life
memorable by calling me daddy. To my wife, Abi, for her love, inspiration,
and unconditional support. To my mom, Lety, and my brother, Jesús, for
their love and always being there.

– Pedro Cesar Santana Mancilla

Contributors

About the authors

Miguel Angel Garcia-Ruiz is an Associate Professor of Computer Science at the School of Computer Science and Technology, Algoma University, Canada. He has taught microcontroller programming and interfacing, human-computer interaction, and interaction design courses. Miguel has a PhD in Computer Science and Artificial Intelligence from Sussex University, England. He has published articles on tinkering with technology applying microcontroller boards. Miguel has conducted research projects funded by Canada's **Northern Ontario Heritage Fund** (**NOHFC**), Algoma University, and the Mexican Ministry of Education.

I want to thank the friends and family, who have been close to me and supported me, especially my wife and my son.

Pedro Cesar Santana Mancilla is a research professor at the School of Telematics at the University of Colima in Mexico. His research interests focus on human-computer interaction, ICT for elderly people, Internet of Things, and machine learning. He is currently serving as president of the **Mexican Association on Human-Computer Interaction** (**AMexIHC**). He is a Senior Member of the IEEE, and ACM and serves as Chair of the Mexican ACM SIGCHI Chapter (CHI-Mexico). Pedro is a member of the **Mexican Academy of Computing** (**AMexComp**) and the **Mexican Society of Computer Science** (**SMCC**).

I want to thank my family and friends, who have always supported me. A special thank you to my wife and son for all their support and patience during the long process of writing this book. Thanks also to Prof. Fermín Estrada for his help in resolving my doubts.

About the reviewers

Darshan Jivrajani is an electronics and communication engineer. He has more than 4 years of experience as an embedded system engineer. In his career, he has completed and has been a part of many projects, including smart switches, smart parking systems, indoor navigation, small computers, TV lifting, and many more. He is familiar with ESP32/8266, PIC, Cypress, NXP, Atmel AVR, Stm32, Raspberry Pi, Beaglebone Black, NRF, BLE4.0/5.0/5.1 microcontroller families. On another side of programming, he is comfortable with embedded C/C++, Python, Node.js, and various communication protocols, such as TCP/IP, UDP, Socket, MQTT, CoAP, I2C, UART, USART, I2S, WiFi, Bluetooth, Zigbee, GPS/GSM, Nb-IoT, Lora, Z-wave, and more.

Paras Balasara is an embedded hardware engineer who creates and brings life to hardware that interacts with sensors through a power supply.

Table of Contents

6
Morse Code SOS Visual Alarm with a Bright LED

7
Creating a Clap Switch

8

Gas Sensor

9

IoT Temperature-Logging System

10

IoT Plant Pot Moisture Sensor

14
COVID-19 20-Second Hand Washing Timer

Other Books You May Enjoy

Index

Preface

This book will introduce you to microcontroller technology. It focuses particularly on two very capable microcontroller boards, the Blue Pill and the Curiosity Nano, and how to connect sensors to them to solve problems and to support everyday life situations. In addition, this book covers the use of **light-emitting diodes (LEDs)** and **liquid-crystal displays (LCDs)** for showing sensor information to its microcontroller board users.

Microcontroller boards are practical small computers used for getting information from an environment using sensors. In this book, each chapter will focus on a specific problem to be solved with microcontroller technology, incorporating the use of practical sensors.

Many people from the intended audience would like to start with a microcontroller-based project but they may not know how to begin with it, including what kind of basic hardware and software tools and electronic components they will need to use. This book will cover that.

A chapter in this book introduces you to the field of electronics, examining and reviewing common electronic components that you will use in this book. Another chapter provides an introduction to C and C++, which will be used for coding Blue Pill and Curiosity Nano applications in most of the chapters.

One of the most important aspects of this book is that sensor programming via microcontroller boards is becoming more effective and easier than before because several easy coding libraries support them, which saves time and effort when getting either analog or digital data from them. This book explains common sensor-programming libraries.

Who this book is for

This book is intended for students, hobbyists, geeks, and engineers alike who wish to dive into the world of microcontroller board programming. In addition, this book is suitable for digital electronics and microcontroller board beginners. If you are already a skilled electronics hobbyist and/or programmer, you may find this book helpful if you want to use and code efficient sensors with microcontroller boards.

People that use other types of microcontroller boards (such as Arduino boards) may find this book useful because it includes an introduction to the Blue Pill and Curiosity Nano microcontroller boards, facilitating the skills transfer required to understand and apply them in electronics projects requiring Arduino microcontroller boards.

Basic knowledge of digital circuits, and C and C++ programming language is desirable but not necessary. This is an introductory book on microcontroller boards for people who are starting with digital electronics projects.

What this book covers

This book covers technical topics on the programming of the Blue Pill and Curiosity Nano microcontroller boards using C++, including descriptions of commonly used sensors and how they are electronically connected to the microcontroller boards. The book consists of 14 chapters, as follows:

Chapter 1, Introduction to Microcontrollers and Microcontroller Boards, introduces the reader to microcontroller technology and explains how to install the **integrated development environments** (**IDEs**) necessary for programming the Blue Pill and Curiosity Nano microcontroller boards that are used in the book.

Chapter 2, Software Setup and C Programming for Microcontroller Boards, provides an overview of C and an introduction to Blue Pill and Curiosity Nano microcontroller board programming, which are used for coding examples in most of the book chapters.

Chapter 3, Turning an LED On or Off Using a Push Button, explains how to use push buttons with microcontroller boards to start a process, such as turning an LED on or off, and how electrical noise from a push button can be minimized.

Chapter 4, Measuring the Amount of Light with a Photoresistor, focuses on how to connect a photoresistor to the Blue Pill and Curiosity Nano microcontroller boards to measure the amount of light within an environment. The result is shown on red, green, and blue LEDs also connected to those boards.

Chapter 5, Humidity and Temperature Measurement, describes how to connect a practical DHT11 sensor to measure the humidity and temperature of an environment, how to display its values on a computer, and also how to use the easy-to-use LM35 temperature sensor, showing its values on two LEDs.

Chapter 6, Morse Code SOS Visual Alarm with a Bright LED, shows how to code the Blue Pill and Curiosity Nano microcontroller boards to display a Morse code SOS signal using a high-intensity LED, increasing its visibility. This chapter also explains how to use a transistor as a switch to increase the LED's brightness.

Chapter 7, Creating a Clap Switch, describes to the reader how to make an electronic wireless control using sounds (claps). When two claps are detected by a microphone connected to a microcontroller board, a signal will be transmitted to activate a device connected to it and an LED will light up.

Chapter 8, Gas Sensor, introduces the reader to the use of a sensor connected to a microcontroller board that reacts with the presence of a specific gas in an environment.

Chapter 9, IoT Temperature-Logging System, shows the reader how to build an **Internet of Things** (**IoT**) temperature logger using the Blue Pill microcontroller board and a temperature sensor. Its data will be transmitted via Wi-Fi using an ESP8266 module.

Chapter 10, IoT Plant Pot Moisture Sensor, explains how to build a digital device with a microcontroller board and a moisture sensor to monitor a plant pot's soil and determine if it needs water, sending an alert wirelessly to notify the user if it's too dry.

Chapter 11, IoT Solar Energy (Voltage) Measurement, continues applying IoT software running on a microcontroller board using the ESP8266 WiFi module to measure voltage obtained from a solar panel through a sensor. The application will send sensor data to the internet using the ESP8266 WiFi signal.

Chapter 12, COVID-19 Digital Body Temperature Measurement (Thermometer), looks at an interesting project to develop a contactless thermometer using an infrared temperature sensor. Its measured temperature data is sent through the I2C protocol to a Blue Pill microcontroller board, displaying it on an I2C LCD.

Chapter 13, COVID-19 Social Distancing Alert, explains how to program a microcontroller board that measures a distance of two meters between two or more people. Within the new normal of COVID-19, we need to maintain social distance due to the higher risk of catching the virus if you are close to someone who is infected. The World Health Organization recommends keeping a distance of at least two meters; this rule varies depending on the country, but it is generally accepted that a distance of two meters is safe.

Chapter 14, COVID-19 20-Second Hand Washing Timer, contains a practical project to make a timer running on a Blue Pill microcontroller board that ensures that people wash their hands for twenty seconds, as per World Health Organization recommendations, to prevent COVID-19 infection. This project shows the time count on a **liquid-crystal display** (**LCD**). An ultrasonic sensor detects if the user is waving at it to initiate the count.

To get the most out of this book

In order to use this book to the full, the reader will need basic knowledge of computer programming and the major operating systems (such as Windows or macOS), although there is a chapter that contains an introduction to C. In order to compile and run the programming examples described in this book, the reader should have the latest Arduino IDE previously installed on their computer (the Blue Pill board can be programmed using the Arduino IDE) and the MPLAB X IDE used for programming the Curiosity Nano microcontroller board; one of the chapters explains how to install and use them. All the program examples contained in this book for the Blue Pill microcontroller board should run on Windows, macOS, and Linux operating systems. The programs that run for the Curiosity Nano microcontroller board were tested on computers running both Windows and Linux operating systems.

If you are using the digital version of this book, we advise you to type the code yourself or access the code via the GitHub repository (link available in the next section). Doing so will help you avoid any potential errors related to the copying and pasting of code.

Some pre-requisites for this book include having basic knowledge of computer programming and electronics, and having some materials, such as a solderless breadboard, many DuPont wires, LEDs, and resistors.

After reading this book, you can continue experimenting with the sensors used in the chapters and perhaps programming and applying other sensors to be connected to microcontroller boards, since this book provides a solid foundation for microcontroller boards programming and use.

Download the example code files

You can download the example code files for this book from GitHub at `https://github.com/PacktPublishing/DIY-Microcontroller-Projects-for-Hobbyists`. In case there is an update to the code, it will be updated on the existing GitHub repository.

We also have other code bundles from our rich catalog of books and videos available at `https://github.com/PacktPublishing/`. Check them out!

Code in Action

Code in Action videos for this book can be viewed at `https://bit.ly/3cZJHQ5`.

Download the color images

We also provide a PDF file that has color images of the screenshots/diagrams used in this book. You can download it here: `https://static.packt-cdn.com/downloads/9781800564138_ColorImages.pdf`.

Conventions used

There are a number of text conventions used throughout this book.

`Code in text`: Indicates code words in text, database table names, folder names, filenames, file extensions, pathnames, dummy URLs, user input, and Twitter handles. Here is an example: "Mount the downloaded `WebStorm-10*.dmg` disk image file as another disk in your system."

A block of code is set as follows:

```
html, body, #map {
  height: 100%;
  margin: 0;
  padding: 0
}
```

When we wish to draw your attention to a particular part of a code block, the relevant lines or items are set in bold:

```
[default]
exten => s,1,Dial(Zap/1|30)
exten => s,2,Voicemail(u100)
exten => s,102,Voicemail(b100)
exten => i,1,Voicemail(s0)
```

Any command-line input or output is written as follows:

```
$ mkdir css
$ cd css
```

> **Tips or important notes**
> Appear like this.

Get in touch

Feedback from our readers is always welcome.

General feedback: If you have questions about any aspect of this book, mention the book title in the subject of your message and email us at customercare@packtpub.com.

Errata: Although we have taken every care to ensure the accuracy of our content, mistakes do happen. If you have found a mistake in this book, we would be grateful if you would report this to us. Please visit www.packtpub.com/support/errata, selecting your book, clicking on the Errata Submission Form link, and entering the details.

Piracy: If you come across any illegal copies of our works in any form on the Internet, we would be grateful if you would provide us with the location address or website name. Please contact us at copyright@packt.com with a link to the material.

If you are interested in becoming an author: If there is a topic that you have expertise in and you are interested in either writing or contributing to a book, please visit authors.packtpub.com.

Share Your Thoughts

Once you've read *DIY Microcontroller Projects for Hobbyists*, we'd love to hear your thoughts! Scan the QR code below to go straight to the Amazon review page for this book and share your feedback.

https://packt.link/r/1-800-56413-9

Your review is important to us and the tech community and will help us make sure we're delivering excellent quality content.

1
Introduction to Microcontrollers and Microcontroller Boards

In this chapter, you will learn how to set up fundamental software tools for programming microcontrollers, as well as how to use basic electronic components as a starting point for programming examples for beginners. We will begin with a general introduction to **microcontrollers** and their definitions, as well as their importance and applications in our everyday lives. We will then go on to give a simplified explanation of the digital and analog electronics necessary for the microcontroller projects carried out in the rest of the chapters. Here, we will also explain the basic equipment that may be used in this book's projects. Finally, we will look at a concise description of the **Blue Pill** and **Curiosity Nano** microcontroller boards to be used throughout this book.

Specifically, we will cover the following main topics:

- Introduction to microcontrollers

- An overview of analog and digital electronics necessary for carrying out the book's projects

- Description of the Blue Pill and Curiosity Nano microcontroller boards

We will also cover how to install the necessary software drivers and the **integrated development environments** (**IDEs**) for programming the Blue Pill and Curiosity Nano. After completing this introductory chapter, you will be able to apply what you have learned regarding the installation of integrated development environments (a type of software tool) to programming the obligatory *Hello World* programs used to make an LED blink. This will run on the Blue Pill and Curiosity Nano using C. Don't worry if you don't know a lot about C programming yet; we have you covered; *Chapter 2, Software Setup and C Programming for Microcontroller Boards*, includes a gentle but concise C programming tutorial.

Technical requirements

The two microcontroller boards described in this book (Blue Pill and Curiosity Nano) can be programmed using different IDEs. An IDE is a programming and debugging software tool that includes a code editor, a compiling environment, debugging options, and so on. Many of the IDEs are also used to upload your compiled program to a microcontroller board via a USB port connection.

These are the IDEs that you will need to install on your computer:

- **Arduino IDE**: This free IDE was originally created for programming Arduino microcontroller boards, but you can also use it for programming the Blue Pill microcontroller board if you install a library for it.

- **MPLAB ® X IDE**: Made by Microchip, the Curiosity Nano manufacturer. This is a free IDE necessary for programming the Curiosity Nano.

We will explain how to install and use those IDEs in this chapter.

The code used in this chapter can be found at the book's GitHub repository here:

https://github.com/PacktPublishing/DIY-Microcontroller-Projects-for-Hobbyists/tree/master/Chapter01

The Code in Action video for this chapter can be found here: https://bit.ly/3zSOg8O

For hardware, you will need the following materials:

- A regular LED light – any color will do.

- A 220-ohm resistor; 0.25 watts.

- A solderless breadboard for connecting an LED and a resistor and some male-to-male jumper wires to make the electrical connections between the components and the microcontroller boards.

- A micro USB cable for connecting your microcontroller boards to a computer.

- The Blue Pill and Curiosity Nano boards, of course! There are several vendors and manufacturers of the Blue Pill board, which uses the STM32F103C8T6 microcontroller. In the case of the Curiosity Nano, we use a version called the *PIC16F15376 Curiosity Nano PIC® MCU 8-Bit Embedded Evaluation Board*, part number DM164148, manufactured by Microchip.

- A programming adapter such as the **ST-Link/V2** is also needed. This electronic interface will allow you to upload the compiled code to the Blue Pill, establishing communication from your computer to the Blue Pill microcontroller board. The ST-Link/V2 needs four female-to-female DuPont wires.

Some of the sensors used in this book can be found in a sensor kit in the form of practical modules, such as the Kumantech 37-in-1 sensor kit:

```
http://www.kumantech.com/kuman-new-version-37-sensor-module-
robot-project-starter-kit-for-arduino-r3-mega2560-mega328-
nano-uno-raspberry-pi-rpi-3-2-model-b-b-k5_p0017.html.
```

This kit can be used with many types of microcontroller boards, including the Blue Pill and the Curiosity Nano. Sometimes, it is convenient to buy a sensor kit like this one for experimenting with its sensor modules. Some other kits include components such as resistors and code examples.

Introduction to microcontrollers

In this section, we will focus on what a microcontroller is and what its main parts are. It is important to understand what the microcontrollers are capable of and how they are used as a fundamental part of many **embedded systems**, so they can be used in real-world projects. An embedded system is a computer subsystem that usually works as part of a larger computer system, for example, a wireless router containing a microcontroller. Let's start with a definition of microcontrollers.

A microcontroller (also known as a microcontroller unit, or MCU) is a very small computer system self-contained in an **integrated circuit** (**IC**). It encases all the necessary computing components to execute tasks, computes numeric calculations, reads data from sensors, keeps data and a program in memory, and send data to actuators, among other actions. Most of the microcontrollers perform **analog-to-digital conversion** (**ADC**), obtaining analog data from sensors and converting it to digital values. More on ADC is explained in *Chapter 4, Measuring the Amount of Light with a Photoresistor*. Digital values are defined by binary values (1 or 0). The next section explains more about those values.

Microcontrollers have an internal clock signal that is like a *heartbeat* that coordinates how tasks and other actions are performed in the microcontroller. This clock signal is not as fast as microprocessors (used by desktop computers and laptops), but it is enough for doing basic operations such as reading a sensor or controlling a motor. Their internal memory is limited, but enough for storing a program capable of running a particular task. In general, microcontrollers do not use an external data storage device such as a hard drive. Everything they need to run is encased in their IC.

An IC is an electronic circuit densely packaged in a small and flat piece of plastic. It contains many microscopic electronic components and electrically connected pins. ICs are manufactured in different packaging. **Dual in-line packaging** (**DIL**) houses two rows of electrically connecting pins. **Quad flat packaging** (**QFP**) includes 8–70 pins per side, useful for surface mounting soldering. Microcontrollers are encased in ICs, as well as other electronic parts.

The pins of some microcontrollers are organized into two rows using DIL packaging. Other ICs, such as the STM32 microcontroller, have four rows of pins, which is known as QFP.

Microcontrollers are also called *a computer in a chip*. They generally have low-power consumption, and, of course, are reduced in size. Some of them are smaller than a fingernail! Microcontrollers are generally used to perform a specific task and execute one particular application, such as controlling the internal functions of a coffee maker, one at a time. Microcontrollers are applied in situations where dedicated and limited computer functions are needed.

Microcontroller boards

A **microcontroller board** is an electronic circuit containing a microcontroller and other supporting components such as voltage dividers/shifters, a USB interface, connection pins, resistors, capacitors, and an external clock.

The purpose of microcontroller boards is to facilitate the connection of external devices, sensors, and actuators to microcontrollers, accelerating project prototyping. For example, the Blue Pill microcontroller board contains its microcontroller at its center, and it has some other components supporting its functions.

Microcontroller boards such as the Blue Pill have **input/output (I/O) ports**, or pins, where sensors, motors, and other electronic components and devices are connected to them. The boards will either read or send data to them through the ports. The boards also have useful pins such as the ground and voltage pins, so sensors and other components can be connected to them to work. Some I/O pins read analog voltages coming from sensors or send analog voltages to actuators (for example, motors), and others are digital pins used for reading and sending digital voltages, typically 0 and 5 volts, or 0 and 3.3 volts. All computers (including microcontrollers) work internally with digital binary numbers containing 0s and 1s. The binary value 0 is represented by 0 volts, and the binary value 1 is represented by either 3.3 or 5 volts. For example, a digital value (1) sent to a digital port could turn on an LED connected to it.

The next section defines what electronics is, and what are analog and digital electronics. These definitions are important in understanding how some electronic components and electronic circuits work, which will be used in this book's chapters.

An overview of analog and digital electronics necessary for carrying out the book's projects

Electronics is the branch of technology and physics concerning the emission and behavior of electrons moving in a conductor, semiconductor, gas, or vacuum. Electronics also deals with the design of electronic circuits and devices. *Figure 1.1* shows a diagram of a basic electronic circuit consisting of a power source (the batteries), a resistor, and a light source (a light-emitting diode, or LED):

Figure 1.1 – An example of an electronic circuit

The electrons flow from the battery's negative (black) terminal through the circuit passing through the LED illuminating it. Don't worry if you don't understand this circuit and its components yet. We will review these in the next paragraphs, and we will use them in other chapters. Analog electronics are electronic circuits that provide and process continuous variable voltage signals, for example, analog voltages that change from 0 to 3.3 volts. Conversely, digital electronics provide and process discrete voltage signals that represent binary values. For example, 0 volts represents a 0 in binary, and 3.3 volts represents a 1 in binary, and no other voltages are used in between. This is how computers and microcontrollers work internally at the lowest level. Microcontrollers convert analog values to digital values internally in order to process incoming signals and then process them digitally. This is called **analog-to-digital conversion** (**ADC**). We will need to understand four key electronics terms that will be covered in other chapters, which are standard units used to measure the flow of electrons:

- **Current**: Current is the rate of flow of electrons in a circuit. Electrons flow through a conductive material from the negative pole of a power source (such as a battery) to its positive pole. This is known as **direct current** (**DC**). The negative side is called ground (GND, or G), sometimes also called earth. Current is measured in amperes or *amps*, denoted by the letter *I* or *i*.

- **Voltage**: This is an electrical measurement of the difference in potential energy between the positive and negative poles of a power source in an electronic circuit. It is measured in volts (V). It is considered as the *pressure* from an electrical circuit's power source *pushing* charged electrons (current) through an electric/electronic circuit.

- **Power**: Power is a rate measurement of how an electric or electronic circuit or device converts energy from one form to another. Power is measured in watts (W). For example, a 60 W lightbulb is brighter than a 40 W lightbulb because the 60 W lightbulb converts electrical energy into light at a higher rate.

- **Resistance**: The electrical resistance of an electrical conductor is the measurement of the difficulty of the electrons in passing an electric current through the conductor. It is measured in ohms, denoted by the Greek letter Omega (Ω). Ohm's law describes the conductivity of many electrically conductive materials. It establishes that the current between two points in a conductor is directly proportional to the voltage across the two points, where its resistance is constant. This law can be mathematically described as I=V/R, and it is very useful for calculating either current, voltage, or resistance in an electronic circuit.

In this section, we have covered fundamental standard measurement units used in electronic circuits that you will apply in all the chapters of this book. The next section deals with important electronic components that you will also need to know before starting experimenting with electronic circuits and microcontroller boards.

Basic electronic components

The following are electronic components commonly used in many microcontroller board projects and in most of the projects described in this book. They allow us to control the current in electronic circuits. We will review four main electronic components: the **resistor**, the **diode**, the **capacitor**, and the **transistor**.

The resistor

Resistors are generally used to reduce the flow of electrons in an electronic circuit. Resistance is useful for allowing some components such as LEDs to work properly in a circuit without burning them. The level of resistance in a resistor can be either *fixed* or *variable*. Some resistors can range from one to thousands of ohms (kilo-ohms or kΩ) to millions of ohms (mega-ohms or MΩ). Resistors are also measured by their power rating measured in watts. This refers to how much current they tolerate without overheating and then failing.

Figure 1.2 shows how to read the values of a resistor:

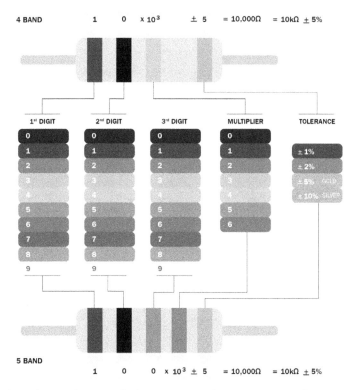

Figure 1.2 – A chart showing how to calculate a resistor value. Image source: "Resistor Color Code", by Adim Kassn, licensed under CC-BY-SA-3.0

Important note

As shown in *Figure 1.2*, the band colors are:

0: Black

1: Brown

2: Red

3: Orange

4: Yellow

5: Green

6: Blue

7: Violet

8: Silver

9: White

The 1% tolerance band is colored brown, the 2% is colored red, the 5% is colored gold, and the 10% is colored silver. You can access the chart in color via this link: `https://commons.wikimedia.org/wiki/File:Resistor_color_code.png`.

Common resistors have four color bands (some have five color bands, but this is rarely used in general electronic circuits) determining their ohm values:

1. The first band indicates the first digit in its ohm value.

2. The second band shows the second digit.

3. The third band indicates the third digit (which is typically the number of zeros).

4. The fourth band determines the resistor tolerance, which is a degree of resistance precision. For example, if the band is colored silver, this means that the resistor will have a 10% tolerance change according to its marked value.

Many resistors used in microcontroller board projects use ¼ watt resistors, which are enough for simple applications.

The diode

The **diode** is an electronic component that allows the flow of current in one direction only. Current in a circuit flows into a diode via its **anode (+)** and flows out through its **cathode (-)**. Diodes are generally used to protect parts of an electronic circuit against reverse current flow. They also help to convert **alternate current** (**AC**) to DC, among other applications. Diodes are also used to protect microcontroller boards when we connect motors to them to avoid voltage *flyback*. This happens when a sudden voltage spike happens across a motor when its supply current is suddenly interrupted or reduced. However, diodes cause a drop in the voltage of around 0.7 V. Diodes are manufactured to handle a certain amount of amperes (current) and voltage. For example, the 1N4004 diode is rated to handle 1 ampere (A) and 400 volts (V), much higher than we will be using in our book's projects. The band around the diode indicates the cathode, generally connected to the ground terminal of a power source. The other pin is the anode, generally connected to the positive (+) terminal of the power source. A common type of diode is the **light-emitting diode (LED)**, which glows when there is a flow of electrons passing through it. They come in different sizes, colors, and shapes. As with regular diodes, LEDs are polarized, so the current enters and leaves the LED in one direction. If too much current passes through the LED, this will damage it. You will need to connect a resistor in series to reduce its current and thus protect it. A resistor with a value of at least 220 ohms should be enough for many microcontroller board applications.

The capacitor

This is an electronic component that temporarily holds (stores) an electric charge. Once the current stops flowing through the capacitor, the charge remains in it and it can be discharged as soon as the capacitor is connected to a circuit. The amount of charge that a capacitor can store is measured in farads (f). Since a farad is a very large amount, many capacitors are made with less than one farad. Capacitors accept certain voltage maximums. 10, 16, 25, and 50 V capacitors are common in microcontroller applications. There are two types: **monolithic** (they don't have polarity) and **electrolytic** (they have polarity).

Electrolytic capacitors are bigger than monolithic capacitors, and their polarity is shown as a band on one side marking the cathode pin and another band marking the anode pin. Remember that the cathode pin is connected to the ground terminal of the power source and the anode is connected to the positive voltage terminal of the power source. Typical values of electrolytic capacitors range from 1 microfarad up to 47,000 microfarads. Capacitors can be used in microcontroller board projects for filtering out (cleaning up) digital or analog signals (removing electrical noise), they can convert alternate voltage to direct voltage, and so on. Be very careful when you're using polarized (electrolytic) capacitors! They can hold lots of energy. You should never touch its legs (pins), short circuit, or connect them in reverse. Make sure you connect an electrolytic capacitor in a project by connecting its positive (+) pin to the positive pole of the circuit's **power supply** (an electronic/electric component that supplies steady power to an electronic circuit or electrical device) and by connecting the capacitor's negative pin to the negative pole of the circuit's power supply. Respect its polarity. If you connect them in reverse (wrong polarity), they will be damaged and can explode. Monolithic (ceramic) capacitors do not have polarity. It doesn't matter how their legs (pins) are connected in the circuit. The typical capacity range of capacitors is from 0.5 picofarads up to 1 microfarad.

The transistor

A transistor can act as a very fast digital switch. Transistors are useful for switching on or off other circuits or devices that require a high current, such as motors and fans. It can also be used as a current amplifier and to form logic gates (**AND, OR, NOT**, and so on); this current is also called a load. Popular and inexpensive examples are the **BC548** and **2N2222** transistors. Transistors are made to hold a certain amount of current and voltage (for example, the BC548 transistor holds a maximum current of 100 mA and 30 V).

Figure 1.3 shows the basic electronic components explained in this section:

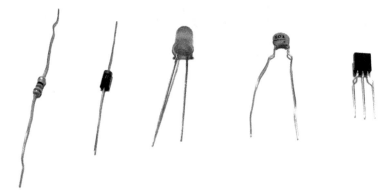

Figure 1.3 – Electronic components (shown from left to right): a resistor, a diode, an LED, a monolithic capacitor, and a transistor

The next section describes a tool called the solderless breadboard, which is very useful for interconnecting electronic components and microcontroller boards.

The solderless breadboard

Another very useful piece that you can use in microcontroller board projects is the solderless breadboard, shown in *Figure 1.4*. It is used for the rapid prototyping of electronic circuits. Its plastic base has rows of electrically connected sockets, coming in many sizes, shapes, and colors:

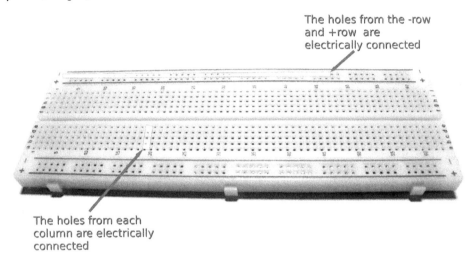

The holes from the -row and +row are electrically connected

The holes from each column are electrically connected

Figure 1.4 – The breadboard's interconnection of columns and rows

Bear in mind that if you insert two wires in one vertical row, then they will be electrically connected. The horizontal rows marked with – and + signs are electrically connected horizontally, as shown in *Figure 1.4*.

This section described important and useful electronic components such as resistors and LEDs, which are commonly used in electronic projects involving microcontroller boards. The next section describes the Blue Pill and Curiosity Nano microcontroller boards used in this book.

Description of the Blue Pill and Curiosity Nano microcontroller boards

This section explains the Blue Pill and Curiosity Nano microcontroller boards, shown in the following photos. The holes from their upper and lower rows will be connected to the header pin, and most of them are the ports. *Figure 1.5* shows the Blue Pill microcontroller board showing the STM32 microcontroller chip at the center:

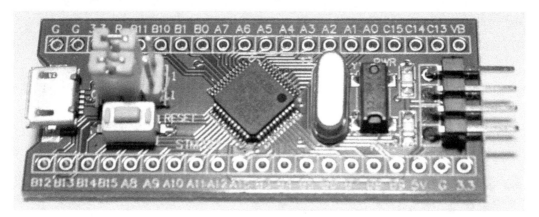

Figure 1.5 – The Blue Pill microcontroller board

Figure 1.6 is a closeup of the Curiosity Nano microcontroller board. Notice its PIC16F15376 microcontroller chip shown at the center:

Figure 1.6 – The Curiosity Nano microcontroller board

The black *rhombus* components at the center of the boards are the microcontrollers. Both boards from the previous photos do not yet have header pins. The reason is that in some projects, it is necessary to solder wires and electronic components directly to a headless board, although in this book you won't need to do that. However, some Curiosity Nano boards allow the pins to be inserted and stay firmly in place without soldering them. In addition, you can buy the Blue Pill with the header pins already soldered. Alternatively, you can solder the header pins to both the Blue Pill and Curiosity Nano boards. Here is a tutorial on how to solder electronics components: `https://www.makerspaces.com/how-to-solder/`.

Figure 1.7 shows the Blue Pill microcontroller board upside down with the header pins already soldered to it and a separate row of pins:

Figure 1.7 – A Blue Pill showing its pins already soldered to it

Once the boards have the header pins in place, you can insert them in solderless breadboards and start prototyping with them without the need for soldering wires or components directly to the boards, which is how we will use them in the projects of this book.

We have selected these microcontroller boards for this book because they are inexpensive, versatile (you can connect different types of sensors, LED lights, motors, and so on, to them), and are reasonably easy to program and use. Microcontroller boards greatly facilitate building prototypes in a short time. You can also apply them in COVID-19-related projects, as we will review this at various points through the book.

In addition, it is always good to learn about boards that use microcontrollers from different families and manufacturers because each one has different capabilities and strengths. The Blue Pill has a microcontroller from the STM32 microcontroller family, and the Curiosity Nano works with a microcontroller from the PIC family.

The Blue Pill has 37 general-purpose I/O pins, including ports PA0 – PA15, PB0 – PB15, and PC13 – PC15. For example, the I/O port PC13 is labeled as *C13* on the Blue Pill.

The Curiosity Nano has 35 GPIO ports, including RA0, R1, RA2, RA3, RA4, RA5, RB0, RB3, RB4, RC7, RD0, RD1, RD2, RD3, RC2, RC3, RB2, RB1, RC4, RC5, RC6, and RD4, among others. We will review the programming of those ports in *Chapter 2, Software Setup and C Programming for Microcontroller Boards*. *Table 1.1* shows technical specifications of the two microcontroller boards used in this book:

Characteristics:	Curiosity Nano:	Blue Pill:
Microcontroller:	PIC16F15376 (8-bit Microchip PIC16 family)	STM32F103C8T6 (32-bit ARM Cortex M3 family)
CPU frequency:	32 MHz	72 MHz
Ports:	35 GPIO (general purpose input/output) ports (pins), 6 analog I/O ports, working at 3.3 volts	37 GPIO (general purpose input/output) ports (pins), 10 analog input pins, working at 3.3 volts
Memory:	16 KB of program memory, 2KB RAM	128 KB of program memory, 20 KB RAM
Operating voltage:	5 volts (obtained from USB port)	5 volts (obtained from USB port)
Serial monitor:	MPLAB-X IDE does not provide a direct way to monitor serial data	Data can be sent to serial (USB) port and displayed by the Arduino IDE
Communication protocols:	I2C, SPI, SMBus, PMBus, EUSART	I2C, SPI, UART, USB, CAN

Table 1.1 – Blue Pill and the Curiosity Nano's technical specifications

Both the Blue Pill and the curiosity Nano run at a much higher speed than most of the Arduino microcontrollers. For example, the Arduino Uno microcontroller board runs at 16 MHz.

Installing the IDEs

Next, we will explain the necessary steps to install and use the IDEs for programming the Curiosity Nano and the Blue Pill microcontroller boards.

Installing the MPLAB X IDE for the Curiosity Nano board

The next steps show how to download and install the MPLAB X tool, which is used for programming the Curiosity Nano. This section also explains the main parts of the MPLAB X IDE:

1. You should sign in first to the free myMicrochip service (Microchip is the manufacturer of the Curiosity Nano). Fill out the registration form on this web page: `https://www.microchip.com/wwwregister/RegisterStep1.aspx`.

2. Once you're registered, download the MPLAB X IDE from this link: `https://www.microchip.com/mplab/mplab-x-ide`.

3. Go to the **Downloads** tab and download the IDE according to the operating system that you are using.

4. Once you have downloaded the installer, follow these instructions for installing the MPLAB X IDE: `https://microchipdeveloper.com/mplabx:installation`.

5. You will also need to download and install the free XC8 C compiler for programming the Curiosity Nano. Open this link: `https://www.microchip.com/mplab/compilers`.

6. Then, go to the **Compiler Downloads** tab and download an installer file according to your operating system.

7. Download the latest version of the XC8 compiler from there. This version is suitable for programming the PIC16F15376 microcontroller that comes with the Curiosity Nano used in this book. Follow the instructions for installing the XC8 compiler here: `https://microchipdeveloper.com/xc8:installation`.

It may take some time to download the XC8 compiler, so be patient. The next section describes commonly used components from the MPLAB X IDE.

Understanding the main components of the MPLAB X IDE

This section describes the main parts of the IDE that you will use to edit your program, compile it, and so on.

Figure 1.8 is a screenshot of the MPLAB X IDE:

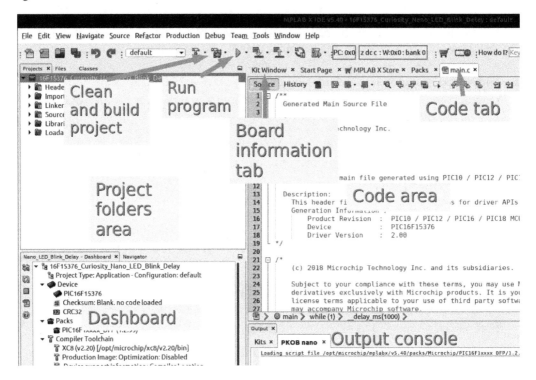

Figure 1.8 – MPLAB X IDE's main parts

The main parts described in *Figure 1.8* include the text editor (code area), which will be used in other chapters of this book. One of the most important buttons in the IDE is the **Run** button, which compiles, runs, and uploads the binary (compiled) file to the Curiosity Nano board.

Here are the steps for starting a new coding project:

1. Click on **File/New Project** to create a new project.

2. Click on the **Code** tab and start writing your code in the code area.

3. Click on the **Run** icon to compile and upload your code to the microcontroller board. The console will show messages on the process and establish whether everything went to plan.

Unlike other microcontroller boards such as the Blue Pill, you need to configure input and output ports in the Curiosity Nano microcontroller board previous to its use. You do this using a special plugin called the **MPLAB X Code Configurator** (**MCC**). You should install the MCC plugin for the MPLAB X IDE. The MCC is a free graphical programming environment that facilitates the configuration of the microcontroller ports, among other applications. It will generate C programming header libraries necessary for reading data from, and writing data to, the microcontroller ports.

This web page explains how to install the MCC in the MPLAB X IDE: `https://www.microchip.com/mplab/mplab-code-configurator`.

The easiest way to install it is by clicking on **Tools/Plugins** on the MPLAB X IDE's main menu, and then downloading and installing it from there.

We have now created an MPLAB X project template where we have configured input and output ports for the Curiosity Nano, as well as handling digital and analog data on them. We already used the MCC plugin to set up the I/O ports so you may no longer need to use the MCC plugin for the projects explained in this book and for other projects. It is a convenient template since it has all the necessary libraries for handling the input and output C-programming functions for some of the Curiosity Nano ports. We will review these libraries and the C-programming functions in *Chapter 2, Software Setup and C Programming for Microcontroller Boards*. The template project is called `16F15376_Curiosity_Nano_IOPorts.zip` and is stored on our GitHub's main page.

Just download the zip file, unzip it, and open the project in MPLAB X.

Installing the Arduino IDE and the Blue Pill library

You can use the Arduino IDE for programming the Blue Pill microcontroller board. Perform the following steps to install the Arduino IDE:

1. Download the Arduino IDE from its official website for Windows, macOS, or Linux from this site: `https://www.arduino.cc/en/main/software`. Don't forget to download the correct IDE installer for your operating system.

2. Run the installer that you just downloaded and follow the onscreen instructions.

3. Identify the Arduino IDE's main options.

Figure 1.9 shows the Arduino IDE and its main parts and areas. The console is a useful component where the IDE shows error or warning messages. The status bar shows the program compilation and uploading status:

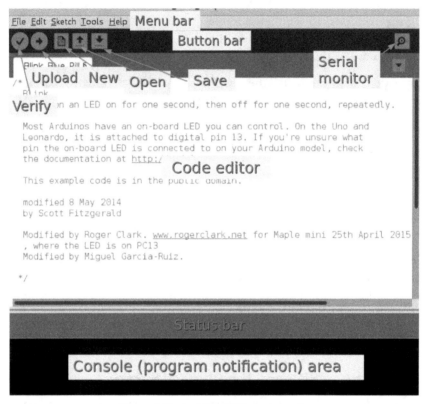

Figure 1.9 – Arduino IDE's main parts

Figure 1.9 shows the main parts of the Arduino IDE, including its code editor, status bar, menu bar, and console. One of the most commonly used features is the **Upload** button, which compiles your program and uploads its compiled code to the Blue Pill. The following steps show how to upload a program to the Blue Pill and how to install a necessary library in the IDE:

1.   Open an example program by clicking on **File/Examples/01.Basics/Blink** from the Arduino's menu bar. This will help you to get familiar with the Arduino IDE. A program called *Blink* will open. Click on the **Verify** button to compile it and generate the binary file for the microcontroller board. The **Upload** button will verify, compile, generate the binary file, and then upload the file to the microcontroller board.

2. Before uploading a program to the Blue Pill, you will need to install a library on the Arduino IDE for programming the Blue Pill. To install the Blue Pill library, click on **File/Preferences** from the IDE's **Menu** bar. Then, a new window will appear. Click on the small window icon to the right of the **Additional Boards Manager URLs** and add this link for installing the library: `http://dan.drown.org/stm32duino/package_STM32duino_index.json`, as shown in *Figure 1.10*:

Figure 1.10 – The IDE's preferences option for writing the link's library

3. Click **OK** in the top window shown in *Figure 1.10* (**Additional Boards Manager URLs**), and then click on the **OK** button in the **Preferences** window.

4. Now, on the **Menu** bar, go to **Tools/Board/Boards Manager**. This will open the **Boards Manager** dialog box.

5. Make sure that you select **All** in the **Type** field. In the **Boards Manager**, search for STM32F1xx, and just install the package that appears on the box.

6. Close that box and click on **Tools/Board/STM32F1 Boards** and then select the **Generic STM32F103C series** option, as shown in the following screenshot. Make sure its variant is **64k Flash**, the upload method is **STLINK**, and its CPU speed is **72MHz, Optimize: Smallest**. *Figure 1.11* shows the Blue Pill configuration in the IDE:

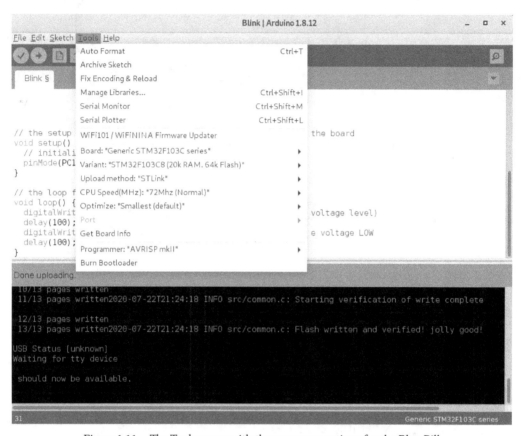

Figure 1.11 – The Tools menu with the necessary options for the Blue Pill

The IDE should be ready to start coding for the Blue Pill. Before that, we need to install an interface called **ST-Link/V2** to flash (upload) our code to the Blue Pill. We cannot upload our compiled program directly to the Blue Pill, as with other microcontroller boards, through the micro-USB cable. One easy way to do this is to use the ST-Link/V2 interface, which is a **Single Wire Interface Module** (**SWIM**) using only four wires.

The ST-Link/V2 is a USB interface that is used for programming and debugging STM32 microcontroller applications and is mainly used for uploading a program to the Blue Pill microcontroller board.

Before connecting the ST-Link/V2, you will need to install its driver on your computer. The following explains how to install the ST-Link/V2 driver for Windows. Download the driver from here: `https://www.st.com/en/development-tools/stsw-link009.html`.

Download and uncompress the zip file and run either `dpinst_amd64.exe` or `dpinst_x86.exe`, depending on whether you are using a 32-bit or 64-bit computer. Most of the recent ones are 64 bits. Follow the displayed instructions on how to install its driver.

Here are the instructions for installing the ST-Link/V2 on macOS: `https://www.st.com/en/development-tools/stsw-link007.html#overview`.

Here are the instructions for installing the ST-Link/V2 on Linux: `https://freeelectron.ro/installing-st-link-v2-to-flash-stm32-targets-on-linux/`.

> **Important note**
>
> If you run the Arduino IDE on Linux, run the IDE as root to get USB access privileges.
>
> Alternatively, you can try this open source toolset for installing the ST-Link/V2 on Windows, macOS, or Linux: `https://github.com/stlink-org/stlink`.

The Arduino IDE provides a serial monitor, accessed by clicking on **Tools/Serial Monitor** from the IDE's main menu. It will show data sent from the Blue Pill to the USB serial port by using special coding functions. *Chapter 5*, *Humidity and Temperature Measurement*, explains how to use the serial monitor. It can be used to show on the computer screen data obtained from sensors, values of variables, and other similar operations.

The next section describes how to run a simple program that will make an LED blink once a second using both the MPLAB and the Arduino IDEs. This will be a practical example to get you familiar with programming the Blue Pill and the Curiosity Nano boards.

Your first project – a blinking LED

This small project demonstrates how to connect an LED to a microcontroller board and how to program one of their I/O ports so that you can turn the LED on, wait for 1 second (1,000 milliseconds), turn the LED off, wait another second, and turn the LED on, in an endless loop.

The project also demonstrates how to upload a compiled program to a microcontroller board. This is an important starter project, since you can later reuse this code for sending a signal to a port and controlling a more complex application, such as a fan. This is like a *Hello World* project for microcontroller boards! We will run this project for both the Blue Pill and the Curiosity Nano microcontroller boards using their respective IDEs.

Running the blinking LED example with the Blue Pill board

This small project demonstrates how to turn an LED on for 1 second, and then off for 1 second, repeatedly. Of course, it also demonstrates how to declare and use an I/O port from the Blue Pill as an output.

> **Tip**
> Before you start, be careful when you manipulate the Blue Pill and Curiosity Nano boards. They can be damaged by your body's static electricity, so you should touch a big metallic area such as a desk frame before manipulating them. You can also wear an anti-static wrist strap. By doing that you are discharging your static electricity. As a general rule, avoid touching the header pins with your bare hands.

We will now look at how to connect the electronic components to the solderless breadboard and the Blue Pill:

1. You will need to insert the Blue Pill board into the breadboard. Do it carefully, since the header pins may bend.

2. Connect the LED to the breadboard. Now, connect one pin of the 220-ohm resistor to the longest pin (leg) of the LED, as shown in the following diagram. Connect the other pin of the resistor to the pin labeled *C13* of the Blue Pill.

3. Connect the shortest pin of the LED to the ground pin labeled *G* or *GND* of the Blue Pill using a wire. Remember that the holes of each column of the breadboard are internally connected.

4. Now, connect the ST-Link/V2 module pins to the Blue Pill pins as follows. The ST-Link/V2 pins are labeled on one of its sides. The Blue Pill's pins are labeled on its bottom side.

5. Connect the Blue Pill's CLK pin to the ST-Link/V2's SWCLK pin.

6. Connect the Blue Pill's DIO pin to the ST-Link/V2's SWDIO pin.

7. Connect the Blue Pill's GND pin to the ST-Lin/V2's GND pin

8. Connect the Blue Pill's 3V3 pin to the ST-Link/V2's 3.3 V pin.

The connections between the ST-Link/V2 and the Blue Pill are shown in *Figure 1.12*:

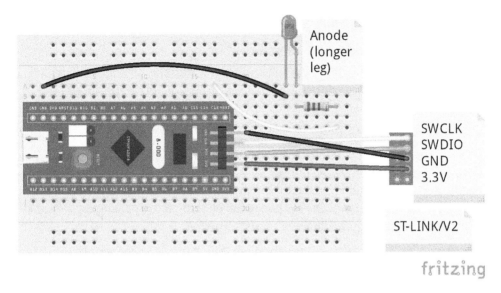

Figure 1.12 – The Blue Pill, LED, and ST-Link/V2 connections

Figure 1.12 shows the connections we have made. Please note that here there are four DuPont wires connected between the Blue Pill and the ST-Link/V2. *Figure 1.13* shows a photo of the connections between the Blue Pill and the ST-Link/V2:

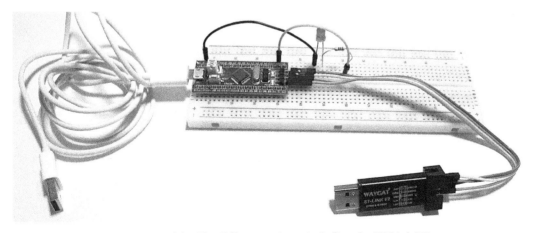

Figure 1.13 – The Blue Pill connections, including the ST-Link/V2

Once you have connected the Blue Pill to the ST-Link/V2, we will continue as follows:

1. Plug in the ST-Link/V2 to your computer. Also, disconnect the micro-USB cable from the Blue Pill (the white cable from the picture). You don't need this for uploading the program to the Blue Pill.

2. Download the program called `Blink_Blue_Pill.ino` from the `Chapter 1` folder located in the book's GitHub repository.

3. Download the `.ino` file from GitHub and open it with the Arduino IDE, which will make a new folder to store the program in. This is a normal practice of the Arduino IDE. Don't worry if you don't understand the code yet. We will explain its C code programming in the next chapter.

4. Click on the upload icon on the IDE to compile and upload the program to the Blue Pill. You will see the Blue Pill's onboard LED and the LED that you connected, blinking once per second if everything went well. The onboard LED is internally connected to the PC13 pin. Try to change the blinking rate by changing the 1,000 millisecond value from the `delay(1000)` function.

5. Once you have uploaded the program to the Blue Pill, it is no longer necessary to keep the ST-Link/V2 connected to it, so you can disconnect it if you want. Now you can connect the Blue Pill's micro USB cable to your computer or a USB power bank. Your compiled program will be kept in Blue Pill's memory and will run every time you power it.

And there you have it! You have completed your first electronic circuit with the Blue Pill. If your LED is blinking, well done!

This project can also be done with Arduino microcontroller boards, such as the Arduino Uno. Just change the port number in the Arduino code and use the Arduino IDE to compile and upload the program. Write `13` Instead of `PC13` in the code for the port number and connect the resistor to digital port number 13 of the Arduino Uno board.

Running the blinking LED example on the Curiosity Nano board

Now let's try the LED blinking example on the Curiosity Nano board. We don't need to connect an interface programmer (such as the ST-Link/V2) to upload your program to the Curiosity Nano, because this board already has the necessary hardware and software components to do so. These are the steps for connecting the Curiosity Nano and the LED:

1. Insert the Curiosity Nano board into the solderless breadboard. Do this carefully because its legs may bend. Also, don't forget to touch a big metallic object before handling the Curiosity Nano to discharge the static electricity from your body.

    Connect the LED to the breadboard. Now, connect one pin of the 220-ohm resistors to the longest pin of the LED, as shown in *Figure 1.14*. Connect the other pin of the resistor to the pin labeled *RE0 pin* on the Curiosity Nano. The RE0 pin is internally connected to the onboard yellow LED from the Curiosity Nano:

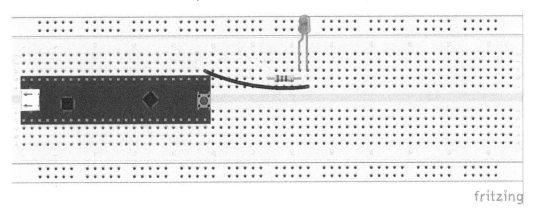

Figure 1.14 – Diagram showing the Curiosity Nano and the LED connection

2. Connect the shortest leg of the LED to the ground pin labeled *GND* of the Curiosity Nano using a wire, as shown in *Figure 1.14*. Any GND pin from the Curiosity Nano will do. Remember that the holes of each breadboard column are internally connected.

The electronic circuit with the Curiosity Nano is shown in *Figure 1.15*. In this diagram, you can visualize with more detail how the LED and the resistor are connected to the board:

Figure 1.15 – The Curiosity Nano and its LED connection

3. Connect the Curiosity Nano to your computer using a micro-USB cable.

4. Now, let's download and run the demo project that contains the C program that makes the LED blink on the Curiosity Nano.

5. Download the file called 16F15376_Curiosity_Nano_LED_Blink_Delay. zip from the Chapter 1 folder located in the book's GitHub repository.

6. Unzip the file and open the project from the MPLAB X IDE. Select the tab with the main.c label. That is the C program that you need to run. Now, click on the run icon (the green triangle) on the MPLAB X.

At this point, the IDE will compile and upload the program to the board. After a few seconds, the onboard yellow LED and the breadboard's LED should be blinking once per second. If this is happening, well done!

Once you connect the LED to the Curiosity Nano and see that the LED is blinking once a second, try to change the code in the IDE a little bit by blinking the LED faster or slower. You can do this by changing the value of the milliseconds from the __delay_ ms(1000); function and don't forget that the value from that function is stated in milliseconds (there are 1,000 milliseconds in a second).

Summary

In this chapter, we have defined what a microcontroller is, as well as its capabilities and limitations. We have also looked at what integrated circuits are (keeping in mind that microcontrollers are a type of integrated circuit) and how their pins are arranged in packages. In addition to this, we analyzed what ports are in microcontroller boards. This is important to know because at some point in future projects, you will need to identify the order of those pins in order to connect sensors or other devices to them. *Table 1.1* showed the hardware description and operating voltage for both the Curiosity Nano and the Blue pill, which is 5 volts.

We then looked at a brief introduction to electronics and the main electronic components used in this and other chapters. We explored how to install two integrated development environment tools for programming the two microcontroller boards used in this book. The two boards have two different ways of uploading a compiled program to them. It is important to compare how two different microcontroller boards work to analyze their capabilities and decide which one you can use in future projects.

Finally, we showcased an initial program on both microcontroller boards that demonstrates how to turn an LED on and off, which can work as a baseline for more complex and detailed projects.

The next chapter contains a concise C programming tutorial, which will be very useful for programming the exercises of the remaining chapters.

Further reading

- Ball, S. (2002). *Embedded Microprocessor Systems: Real-World Design.* Burlington, MA: Newnes/Elsevier Science.

- Gay, W. (2018). *Beginning STM32: Developing with FreeRTOS, libopencm3, and GCC.* St. Catharines: Apress.

- Horowitz, P., Hill, W. (2015). *The Art of Electronics.* [3rd ed.] Cambridge University Press: New York, NY.

- Microchip (2019). *PIC16F15376 Curiosity Nano Hardware User Guide.* Microchip Technology, Inc. Available from: `http://ww1.microchip.com/downloads/en/DeviceDoc/50002900B.pdf`.

- Mims, F.M. (2000). *Getting Started in Electronics.* Lincolnwood, IL: Master Publishing, Inc.

2
Software Setup and C Programming for Microcontroller Boards

In this chapter, you will review the basic configuration of the IDEs used for programming the Blue Pill and Curiosity Nano microcontroller boards, as well as learn the basics of the C programming language necessary for coding applications for the Blue Pill and the Curiosity Nano. This is by no means a comprehensive C tutorial. It contains important information to understand and complete the exercises explained in all the chapters of this book. In this chapter, we're going to cover the following main topics:

- Introducing the C programming language
- Introducing Curiosity Nano microcontroller board programming
- Introducing Blue Pill microcontroller board programming
- Example – Programming and using the microcontroller board's internal LED

By the end of this chapter, you will have received a solid introduction to the C programming language, including a set of programming instructions useful for developing many small and mid-sized microcontroller projects with the Blue Pill and Curiosity Nano microcontroller boards. This chapter also covers the use of the internal LED, which both the Blue Pill and the Curiosity Nano have. This can be very useful for quickly showing digital results (for example, confirming actions in your project).

Technical requirements

The software that we will use in this chapter is the Arduino and MPLAB X IDEs for programming the Blue Pill and the Curiosity Nano, respectively. Their installation process was described in *Chapter 1, Introduction to Microcontrollers and Microcontroller Boards*. We will also use the same code examples that were used in the aforementioned chapter.

In this chapter, we will also use the following hardware:

- A solderless breadboard.
- The Blue Pill and Curiosity Nano microcontroller boards.
- A micro USB cable for connecting your microcontroller boards to a computer.
- The ST-LINK/V2 electronic interface needed to upload the compiled code to the Blue Pill. Remember that the ST-Link/V2 requires four female-to-female DuPont wires.

These are fundamental hardware components that will suffice for the examples described in this chapter, and will also prove useful in other more complex projects explained in other chapters.

The code used in this chapter can be found at the book's GitHub repository here:

`https://github.com/PacktPublishing/DIY-Microcontroller-Projects-for-Hobbyists/tree/master/Chapter02`

The Code in Action video for this chapter can be found here: `https://bit.ly/3xwFvPA`

The next section explains a concise introduction to the C programming language.

Introducing the C programming language

The **C programming language** was initially created in the early seventies for developing the UNIX operating system, but it has been ported to practically all operating systems ever since. It is a mid-level programming language because it shares properties from high-level languages such as Python and low-level languages, for example, the assembly language. The C language is generally easier to program than low-level languages because it is very *human-readable* and there are many libraries available that facilitate the development of software applications, among other reasons. It is also very efficient for programming embedded systems. C is one of the most popular coding languages, and virtually all microcontrollers can be programmed with C compilers – Blue Pill and Curiosity Nano are no exceptions.

The C language is not completely portable among different families and manufacturers of microcontrollers. For example, the I/O ports and the interrupts are not programmed the same in both Blue Pill and Curiosity Nano. That is why two types of C compilers and different libraries are needed for programming both microcontroller boards. In fact, the Arduino IDE used for programming the Blue Pill uses a variant of C called **C++**. C++ is a powerful extension of the C programming language that incorporates features such as object-oriented and low-memory level programming.

The following section explains the basics of the C language structure. This section includes an explanation of the `#include` directive, writing comments, understanding variables, using constants, a keywords list, declaring functions, evaluating expressions, and writing loops in C.

The basic structure of the C language

As with other programming languages, C makes it possible to declare program elements such as constants, types, functions, and variables in separate files called **header files**, which end in .h. This can help to organize C instructions and reduce clutter in your main C code. A library is a header file containing program elements (such as functions) that can be shared with other C programmers or constantly used in different C programs. C language compilers contain important libraries that we will use in this book. The header files can be included (that is, linked and compiled) along your main program using the `#include` directive; hence, the programming elements declared in the header file will be called and used in your C program.

There are many useful standard and non-standard libraries. We will review and use both. The #include directive is a special instruction for the C compiler and not a regular C instruction. It should be written at the beginning of the program and without a semicolon at the end. Only the C statements have a semicolon at the end. There are three ways to write and apply the #include directive. These are as follows:

- #include <file_name.h>: This type of directive uses the less than and greater than symbols, meaning that the header file (.h) is placed in the compiler path. You don't need to write the complete path to the header file.

- #include "file_name.h": This type of directive uses double quotes. The header file is stored in the project's directory.

- #include "sub_directory_name/file_name.h": This directive type tells the compiler that the header file is placed in a sub-directory. Please note that the slash symbol is applied depending on the operating system that you are using. For example, Windows computers use a backslash (\) symbol as a directory separator. Linux and Mac computers use the forward-slash (/) symbol.

The next sub-section shows how to define and use header files.

Example of the #include directive

The following program example shows how to include a header file that is placed in the project's directory:

```
#include "main_file.h"
int main(void)
{
    x = 1;
    y = 2;
    z = x+y;
}
```

In the preceding example, the x, y, and z variables were declared in the main_file.h header file, so they are not declared in the main program. The header file (file.h) contains the following code declaring the three variables used in the main code:

```
int x;
int y;
int z;
```

We could declare the variables in the main program and not declare the variables in a header file (.h). It is up to you whether you want to write program elements in header files. We will learn more about variables later in this chapter.

Note

The C language is case sensitive, so be careful when writing C code. Most C language instructions are written in non-capitalized letters. Be careful when you declare a variable, too. For example, the variables x and X are different in C.

There are standard libraries that come with the C language and many programmers make good use of them. The `stdio.h` library (stored as a header file) is widely used in C programming. It defines several macros, variable types, and also specialized functions for performing data input and output; for example, taking input letters from a keyboard or writing text to the console. The console is a text-based area provided by the IDE where reading data from a keyboard or writing text or special characters happens.

This is a short C program example using the `<stdio.h>` directive:

```
// program file name: helloworld.c
#include <stdio.h>
int main()
{   // start main block of instructions
    printf("Hello world!");
    return 0;
}
```

C program files are stored with the .c extension (such as `mainprogram.c`). The C++ program files are generally stored with the .cpp extension (for example, `mainprogram.cpp`).

The **American National Standards Institute (ANSI)** C language defines a number of useful built-in functions, such as `printf()`, which displays characters (for example, a text message) on the IDE's console. As you can see from the preceding program example, we wrote some comments explaining each line of code. The next section shows the different ways of writing comments in the C language.

Using comments in C

Comments are either blocks or lines of text that don't affect the functioning of C programs. Writing comments in C programming is useful because they can be used to explain and clarify the meaning or the functioning of instructions, functions, variables, and so on. All the comments that we write in the program are ignored by the compiler. There are a couple of ways of writing comments in C:

- Using double slashes (//): This makes a single-line comment.

- Using slashes and asterisks (/*   */): This makes a comment with a block of text.

This code example demonstrates how to use both types of comments:

```
/************************************************************
Program: Helloworld.c
Purpose: It shows the text "Hello, world!!" on the IDE's
console.
Author: M. Garcia.
Program creation date: September 9, 2020.
Program version: 1.0
*************************************************************/
#include <stdio.h>   //standard I/O library
int main(void)
{
    int x; // we declare an integer variable

    printf("Hello, world!!");
    x=1; // we assign the value of 1 to variable x.
}
```

> **Tip**
> It is a good programming practice to write the code's purpose, version number and date, and the author's name(s) as comments at the beginning of your C program.

The next section describes how to declare and use variables in C programming. Variables are very useful, and you will use them in most of the chapters of this book.

Understanding variables in C

A **variable** is a name (also called an identifier) assigned via programming to a microcontroller memory storage area that holds data temporarily. There are specific types of variables in C that hold different types of data. The variable types determine the layout and size of the variable's assigned microcontroller memory (generally, its internal random-access memory or RAM).

We must declare a variable in the C language first to use it in your code. The variable declaration has two parts – a data type and an identifier, using this syntax: `<data_type>` `<identifier>`. The following explains both:

- A **data type** (or just type) defines the type of data to be stored in the variable (for example, an integer number). There are many data types and their modifiers. The following table describes the four main types:

Type	Description	Bit size	Format specifier (used in functions such as printf();)
char	It contains a character and is an integer type (it holds a number), with a range of -128 to 127. The number encodes the ASCII character value.	8	%c
int	Unsigned integer type. It can hold numbers with a range of -32,767 to +32,767.	16	%i or %d
float	It is a real, floating-point type, with a range of 3.4E-38 to 3.4E+38	32	%f
double	It is a real, floating-point type with double precision with a range of 1.7E-308 to 1.7E+308	64	%lf

Table 2.1 – The main four data types used in the C language

- Each type from *Table 2.1* has the modifiers `unsigned`, `signed`, `short`, and `long`, among others. For example, we can declare a variable that holds unsigned integers as `unsigned int x;`.

- There is another type named `void`. This type has no value and is generally used to define a function type that returns nothing.

- An identifier is a unique name identifying the variable. Identifiers can be written with the letters a..z or A..Z, the numbers 0..9, and the underscore character: _. The identifier must not have spaces, and the first character must not be a number. Remember that identifiers are case-sensitive. In addition, an identifier should have fewer than 32 characters according to the ANSI C standard.

For example, let's declare a variable named x that can hold a floating-point number:

```
float x;
```

In the preceding line of code example, the C compiler will assign variable *x* a particular memory allocation holding only floating-point numbers.

Now, let's use that variable in the following line of code:

```
x=1.10;
```

As you can see, we store the floating-point value of 1.10 in the variable named *x*. The following example demonstrates how to use a variable in a C program:

```
/* program that converts from Fahrenheit degrees to Celsius
degrees. Written by Miguel Garcia-Ruiz. Version 1.0. Date:
Sept. 9, 2020
*/
#include <stdio.h> // standard I/O library to write text
int main(void) // It won't return any value
{
    float celsius_degrees;
    float fahrenheit_degrees=75.0;
    // Calculate the conversion:
    celsius_degrees=(fahrenheit_degrees-32)*5/9;
    // printf displays the result on the console:
    printf("%f",celsius_degrees);
}
```

You can initialize a variable with a value when it is declared, as shown in the preceding example for the `fahrenheit_degrees` variable.

We can also store strings in a variable using double quotes at the beginning and end of the string. Here's an example:

```
char   name = "Michael";
```

The preceding example shows how a string is stored in a char variable type, which is an array of characters.

Declaring local and global variables

There are two types of variables in C depending on where they are declared. They can have different values and purposes:

- **Global variables**: These are declared outside all the functions from your code. These variables can be used in any function and through the whole program.

- **Local variables**: Local variables are declared inside a function. They only work inside the function that were declared, so their value cannot be used outside that function. Have a look at this example containing both global and local variables:

```c
#include<stdio.h>
// These are global variables:
int y;
int m;
int x;
int b;
int straight_line_equation() {
    y=m*x+b;
    return y;
}
int main(){
    int answer;   // this is a local variable
    m=2;
    x=3;
    b=5;
    answer = straight_line_equation();
    printf(" %d\n   ",answer);
    return 0;   // this terminates  program
}
```

In the preceding example, the global variables *y*, *m*, *x*, and *b* work in all programs, including inside the `straight_line_equation()` function.

Using constants

Constants (also called constant variables) can be used to define a variable that has a value that does not change throughout the entire program. Constants in C are useful for defining mathematical constants. This is the syntax for declaring a constant:

```c
const <data_type> <identifier>=<value>;
```

Here, the data type can be either `int`, `float`, `char`, or `double`, or their modifiers, for example:

```c
const float euler_constant=2.7183;
const char A_character='a';
```

You can also declare variables using the `#define` directive. It is written at the beginning of a program, right after the `#include` directive, without a semicolon at the end of the line, using this syntax: `#define <identifier> <value>`.

We don't need to declare the constant's data type. The compiler will determine that dynamically. The following examples show how to declare constants:

```
#define PI 3.1416
#define value1 11
#define char_Val 'z'
```

The next section deals with keywords from the C language that are widely used in C programs.

Applying keywords

The ANSI C standard defines a number of **keywords** that have a specific purpose in C programming. These keywords cannot be used to name variables or constants. These are the keywords (statements) that you can use in your C code:

```
auto, break, case, char, const, continue, default, do,
double, else,  enum, extern, float, for, goto, if, int,
long, register, return, short, signed sizeof, static, struct,
switch, typedef, union, unsigned, void, volatile, while.
```

The compilers used to compile programs for the Blue Pill and Curiosity Nano boards have additional keywords. We will list them in this chapter. The following section explains what functions in C are.

Declaring functions in C

A **function** is a block that contains a group of C instructions, or just a single instruction, that will perform a task, or a number of tasks, with a particular purpose. The block of instructions is encompassed with curly brackets. A C program has at least one function, which is `main()`. This function is written in C programs, and other functions are called from it. You can logically divide your code up into functions to make it more readable and to group instructions that are related to the same task, giving the instructions some structure. Functions in C are defined more or less like algebraic functions where you have a function name, a function definition, and a function parameter(s).

The general form for defining a function in C is the following:

```
<return data type> <function name> (parameter list) {
    <list of instructions>
    return <expression>; //optional
}
```

The **parameters** (also called **arguments**) input data to a function. The parameters are optional since a function cannot have parameters at all. We need to declare the type of each parameter. The variables declared as parameters behave like local variables for the function where they were declared. The variables declared as parameters are also called the function's formal parameters, and these are also termed *call by value*, meaning that the changes made to the parameters (the variables) inside their function do not have any effect on them. In other words, the instructions from inside a function cannot alter the values of its function's parameters. The `return` statement allows a value from a function to be returned, and this returned value is used in other parts of the program. The return statement is optional since you can code a function that does not return a value.

> **Tip**
> It is a good programming practice to indent the instructions contained in a function block. This gives the function more visual structure and readability.

The following function example shows how to use parameters and how data is returned from a function, where `number1` and `number2` are the function parameters:

```
int maxnumber(int number1, int number2) {
    /* Declaring a local variable to store the result: */
    int result1;
    if (number1 > number2)
        result1 = number1;
    else
        result1 = number2;
    return result1;
}
```

In the preceding example, the function returns the results of the comparison between the two numbers.

> **Tip**
> Make sure that the function's data type has the same type as the variable used in the `return` statement.

If, for some reason, you don't need to return a value from a function, you can use the `void` statement instead of defining the function's data type, for example:

```
void error_message ()
{
    printf("Error.");
}
```

In the preceding example, we are not using the `return   0` statement in the function because it's not returning any value. We can then **call** the function by its name: `error_message();`.

Calling a function

Once we declare a function, we need to **call** it, that is, run it in another part of your code. This transfers the program control to the called function and it will run the instruction(s) contained in it. After executing all the instructions from the function, the program control resumes, running instructions from the main program.

To call a function, you will need to write the function name and the required values for the parameters. If your function returns a value, you can store it in a variable. For example, let's call the `max()` function that we explained previously:

```
int result2;
result2=maxnumber(4,3);
```

In this example, the result of the number comparison made by the `maxnumber()` function will be stored in the `result2` variable.

Evaluating expressions (decision statements)

The C language provides a way to declare one or more **logic conditions** that can be evaluated (tested) by the program, as well as some statements that need to be executed according to the result of that evaluation, that is, if the condition is either true or false.

The C programming language assumes that the true value is any non-null or non-zero value. It is false if the value is zero or null. C has the following decision-making statements:

- `if (expression_to_evaluate) {statements}`: This has a Boolean expression in the decision that is followed by one or more statements to be run if the decision is true, for example:

```
#include <stdio.h>

void main(){
        int x;
        x=11;

        if (x>10) {
                printf("yes, x is greater than 10");
        }
}
```

- `if (decision) {statements} else {statements}`: The `else` component can be used after an `if` statement and can be useful when running one or more statements if the decision is false, for example:

```
#include <stdio.h>

void main(){
    int x;
    x=5;
    if (x>10) {
        printf("yes, x is greater than 10");
    }
    else {
        printf("no, x is not greater than 10");
    }
}
```

In the preceding example, the x variable is analyzed, and if x is greater than 10, it will print out this message on the IDE's console: `yes, x is greater than 10`, otherwise it will print out `no, x is not greater than 10`.

> **Tip**
>
> Be careful when you evaluate two variables with the `if` statement. Use double equal signs for that (==). If you use only one equal sign, the compiler will raise an error. Do it like this: `if` (x==y) {statements}

- The `switch` statement compares the value of a variable against a number of possible values, which are called cases. Each case from the `switch` statement has a unique name (identifier). If a match is not found in the list of cases, then the default statement will be executed and the program control goes out of the `switch` with the list of cases. The optional `break` statement is used to terminate the program control outside of the `switch` block. This is useful if, for some reason, you don't want the `switch` statement to keep evaluating the rest of the cases. The following is the syntax for the `switch` statement:

```
switch( expression_to_evaluate)
{
    case value1:
        <statement(s)>;
        break;
    case value_n:
        <statement(s)>;
        break;
}
```

The preceding code shows the syntax for the `switch` statement, including its break sentence. The following code is an example of using `switch`, which will compare the variable age against three cases. In case the variable has a value of 10, it will print out the following text: `the person is a child`:

```
#include <stdio.h>
void main(){
    int age;
    age=10;
    switch (age)
    {
        case 10:
            printf ("the person is a child");
            break;
        case 30:
```

```
                    printf ("the person is an adult");
                    break;
            case 80:
                    printf ("the person is a senior citizen");
                    break;
        }
    }
```

So far, we have reviewed how to logically evaluate an expression. The next section explains how to run one or more statements repeatedly. This can be useful for some repetitive tasks for the microcontroller board, such as reading data from an input microcontroller port continuously.

Understanding loops

A **loop** in C executes a single line of code or a block of lines of code a number of times, and if programmed, it could run endlessly. The C programming language provides three ways of making loops using the keywords for, while, and do..while:

for loop

The for loop repeats one or more statements contained in its block until a test expression becomes false. This is the syntax of the for loop:

```
for (<initialization_variable>;
        <test_expression_with_variable>; <update_variable>)
{
    <statement(s)_to_run>;
}
```

In the preceding syntax, the counter_variable initialization is executed once. Then, the expression is evaluated with counter_variable. If the tested expression is false, the loop is terminated. If the evaluated expression is true, the block statement(s) are executed, and counter_variable is updated. counter_variable is a local variable that only works in the for loop. This example prints out a list of numbers from 1 to 10 on the IDE's console:

```
for (int x=1; x<=10; x++)
{
    printf ("%d ", x);
}
```

Please note that the x++ statement is the same as writing x=x+1.

while loop

The while loop repeats one or more statements from its block while a given condition is true, testing its condition prior to executing the statements. When its condition tests false, the loop terminates. Here is the syntax for the while loop statement:

```
while (<test_expression>)
{
    statement(s);
}
```

The preceding code is the syntax for the while loop. The following is example code that uses the while loop, counting from 0 to 10:

```
int x = 0;
while (x <= 10)
{
    // \n it will display the next number in a new
    // line of text:
    printf("%d \n", x);
    x=x+1;
}
```

do..while loop:

This type of loop is very similar to the while loop. The do..while loop executes its block statement(s) at least once. The expression is evaluated at the end of the block. The process continues until the evaluated expression is false.

The following is the syntax for the do..while loop:

```
do
{
    statement(s);
}
while (<test_expression>);
```

The following example uses the do..while loop, counting numbers from 5 to 50, while the sum is < 50:

```
int number=5;
do
{
    number=number+5;
    printf("%d ", number);
}
while (number < 50);
```

In the preceding code, the variable called number has the value 5 added to it and the variable is printed out on the IDE's console at least once, and then the variable is evaluated.

The infinite loops

You can also program an **infinite loop**, which, of course, will run endlessly (the loop does not terminate) until we abort the program (or disconnect the power from the microcontroller board!). Infinite loops can be useful for showing a result from a microcontroller continuously, reading data from a microcontroller board continuously without stopping it, and so on.

You can do this using any of the three types of loops. The following are some examples of infinite loops:

```
for(; ;)
{
    printf("this text will be displayed endlessly!");
}

while(1)
{
    printf("this text will be displayed endlessly!");
}

do
{
    printf("this text will be displayed endlessly!");
}
while (1);
```

As you can see from the preceding code, programming endless loops is easy and simple.

The break and continue keywords in loops

You can **break** running a block of statements from a loop using the break keyword. The break statement of the following example will stop the for loop, but the statement will run only once:

```
for (int x=1; x<=10; x++)
{
    printf("%d ", x);
    break;
}
```

You can use the break statement in any of the three types of loops.

The **continue** keyword will force the next iteration of the loop to run, skipping any statement written after the continue keyword. This example will not print out the second line of text:

```
for (int x=1; x<=10; x++)
{
    printf("%d ", x);
    continue;
    printf("this line won't be displayed.");
}
```

The preceding code displays the value of x without displaying the next line of text because of the continue statement, moving the program control to the beginning of the for loop.

The next section deals with a number of C statements and functions that were created specifically for the Curiosity Nano microcontroller board and that are slightly different from those for the Blue Pill board.

Introducing Curiosity Nano microcontroller board programming

As you learned from *Chapter 1*, *Introduction to Microcontrollers and Microcontroller Boards*, the Curiosity Nano can be programmed using ANSI C language, explained in this chapter, using the MPLAB X IDE.

The basic structure of a C program for the Curiosity Nano is similar to the one explained above using the `main()` function, but its declaration changes. You have to include the keyword void in it, as follows:

```
//necessary IDE's library defining input-output ports:
#include "mcc_generated_files/mcc.h"
void main(void) //main program function
{
    // statements
}
```

The file `16F15376_Curiosity_Nano_IOPorts.zip` from the book's GitHub page contains the necessary **input-output (I/O)** functions for the Curiosity Nano to work. Each port's I/O functions contain the port name. For example, the `IO_RD1_GetValue()` function will read an analog value from the Curiosity Nano's RD1 port.

The following are useful functions that you can use for programming the Curiosity Nano, which is already defined by the MPLAB X compiler. Note that xxx means the Curiosity Nano's port name. Please read *Chapter 1, Introduction to Microcontrollers and Microcontroller Boards*, to familiarize yourself with the Curiosity Nano's I/O port names and their respective chip pins:

- `IO_xxx_SetHigh();`: This function writes the logic HIGH (3.3 V) value on the specified pin (port).

- `IO_xxx_SetLow();`: This function writes the logic LOW (0 V) value on the specified pin (port).

- `IO_xxx_GetValue();`: This function returns the logic (digital) value (either HIGH or LOW) that is read from the specified port. HIGH is returned as 1. LOW is returned as 0.

- `ADC_GetConversion(xxx);`: This function reads an analog value from the specified port and returns a value from 0 to 1023 corresponding to the analog-to-digital conversion done on the read value.

- `SYSTEM_Initialize();`: This function initializes the microcontroller ports.

- `__delay_ms(number_milliseconds);`: This function pauses the program for a number of milliseconds (there are 1,000 milliseconds in one second).

- `IO_xxx_Toggle();`: This function toggles the port's value to its opposite state of the specified port. If the port has a logic of HIGH (1), this function will toggle it to 0, and vice versa.

We will use some of the preceding functions in an example explained later in this chapter.

Figure 2.1 shows the Curiosity Nano's pins. Bear in mind that many of them are I/O ports:

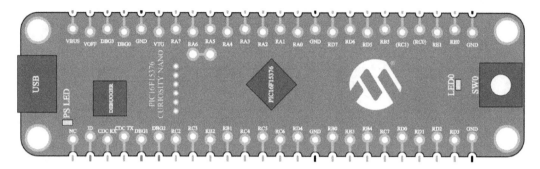

Figure 2.1 – Curiosity Nano's pins configuration

We have configured the following ports from the Curiosity Nano microcontroller board as I/O ports. We did this in all the Curiosity Nano's software project files from this book. The ports' pins can be seen in *Figure 2.1*. Some of them are used throughout this book:

RA0, RA1, RA2, RA3, RA4, RA5, RB0, RB3, RB4, RB5, RC0, RC1, RC7, RD0, RD1, RD2, RD3, RD5, RD6, RD7, RE0, RE1, and SW0.

The following section explains the basic programming structure and important functions for the Blue Pill board microcontroller board coding, which are somewhat different from the Curiosity Nano board.

Introducing Blue Pill microcontroller board programming

As you learned from *Chapter 1, Introduction to Microcontrollers and Microcontroller Boards*, you can program the Blue Pill board using the Arduino IDE, along with a special library installed in the IDE. Remember that this IDE uses C++ language, which is an extension of C. Programs are also called sketches in Arduino IDE programming. All the sketches must have two functions, called `setup()` and `loop()`.

The `setup()` function is used to define variables, define input or output ports (board pins), define and open a serial port, and so on, and this function will run only once. It must be declared before the `loop()` function.

The `loop()` function is the main block of your code and will run the main statements of your program. This `loop()` function will run repeatedly and endlessly. Sketches do not require the `main()` function.

This is the main structure for your sketches (programs):

```
void setup()
{
    statement(s);
}
void loop()
{
    statement(s);
}
```

Here is how to define pins (a microcontroller board's ports) either as inputs or outputs:

```
void setup ( )
{
  // it sets the pin as output.
    pinMode (pin_number1, OUTPUT);

  // it sets the pin as input
    pinMode (pin_number2, INPUT);
}
```

An input port will serve to read data from a sensor or switch, and an output port will be used to send data to another device or component, turn on an LED, and suchlike.

Tip

Programming in the Arduino IDE is case-sensitive. Be careful when you write function names, define variables, and so on.

As you can see from the preceding code, each block of statements is enclosed in curly brackets, and each statement ends with a semicolon, similar to ANSI C. These are useful functions that can be used for programming the Blue Pill:

- `digitalWrite(pin_number, value);`: This function writes a HIGH (3.3 V) or LOW (0 V) value on the specified pin (port); for example, `digitalWrite(13,HIGH);` will send a HIGH value to pin (port) number 13.

> **Note**
>
> You must previously declare `pin_number` as OUTPUT in the `setup()` function.

- `digitalRead(pin_number);`: This function returns either a logic HIGH (3.3 V) or logic LOW (0 V) value that is read from a specified pin (port), for example, `val = digitalRead(pin_number);`.

> **Note**
>
> You must previously declare `pin_number` as INPUT in the `setup()` function.

- `analogWrite(pin_number, value);`: This function writes (sends) an analog value (0..65535) to a specified PIN (output port) of the Blue Pill.
- `analogRead(pin_number);`: This function returns an analog value read from the specified PIN. The Blue Pill has 10 channels (ports or pins that can be used as analog inputs) with a 12-bit **analog-to-digital conversion (ADC)** resolution. Their input range is 0 V – 3.3 V. So, the `analogRead()` function will map input voltages between 0 and 3.3 volts into integer numbers between 0 and 4095, for example:
- `int val = analogRead(A7);`
- `delay(number_of_milliseconds);`: This function pauses the program for the specified amount of time defined in milliseconds (remember that there are one thousand milliseconds in a second).

> **Tip**
>
> You can also use the C language structure explained in this section for programming the Arduino microcontroller boards, with the only difference being that the range of values for `analogWrite()` will be 0...255 instead of 0...65535, and `analogRead()` will have a range of 0 to 1023 instead of 0 to 4095.

Figure 2.2 shows the I/O ports and other pins from the Blue Pill:

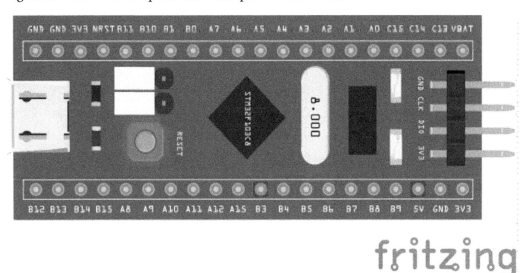

Figure 2.2 – The Blue Pill's pins configuration

The ports' pins can be seen in *Figure 2.2*. Some of them are used in this book's chapters. The Blue Pill has the following analog ports: A0, A1, A2, A3, A4, A5, A6, A7, B0, and B1. The following are digital I/O ports: C13, C14, C15, B10, B11, B12, B13, B14, B15, A8, A9, A10, A11, A12, A15, B3, B4, B5, B6, B7, B8, and B9.

Just remember that in the code, the ports are referenced as PA0, PA1, and so on, adding a letter P.

We will use some of the preceding functions in an example in the next section.

Example – Programming and using the microcontroller board's internal LED

In this section, we will use common statements from C/C++ languages for controlling an internal LED from the Blue Pill and the Curiosity Nano boards. The internal LED can be very useful for quickly verifying the state of I/O ports, showing data from sensors, and so on, without the need to connect an LED with its respective resistor to a port. The next section will show how to compile and send a piece of code to the microcontroller boards using their internal LED.

Programming the Blue Pill's internal LED

This section covers the steps for programming the internal LED. You don't need to connect any external electronic component, such as external LEDs. Using the internal LED from the Blue Pill is useful for quickly testing out and showing the result or variable value from a program. You will only need to use the microcontroller boards. The following steps demonstrate how to upload and run the program to the Blue Pill:

1. Connect the ST-LINK/V2 interface to the Blue Pill, as explained in *Chapter 1, Introduction to Microcontrollers and Microcontroller Boards*.

2. Connect the USB cable to the Blue Pill and your computer. Insert the Blue Pill into the solderless breadboard. *Figure 2.3* shows the internal LED from the Curiosity Nano and the Blue Pill boards:

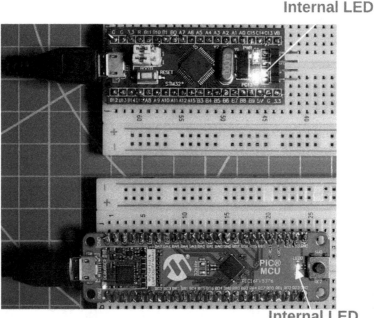

Figure 2.3 – The Blue Pill (top) and the Curiosity Nano's internal LEDs

3. Open Arduino IDE. Write the following program in its editor:

```
/*
Blink
This program turns on the Blue Pill's internal LED
on for one second, then off for two seconds,
repeatedly.
```

```
    Version number: 1.
    Date: Sept. 18, 2020.
    Note: the internal LED is internally connected to
    port PC13.
    Written by Miguel Garcia-Ruiz.
  */

void setup()
{
  pinMode(PC13, OUTPUT);
}

void loop()
{
  digitalWrite(PC13, HIGH);
  delay(1000);
  digitalWrite(PC13, LOW);
  delay(2000);                  // it waits for two seconds
}
```

4. Click on the **Upload** button from the IDE and see how the program is compiled and sent to the Blue Pill. Once it's done, you should see the Blue Pill's small LED blinking.

> Tip
> You can use the preceding code for blinking the internal LED of the Arduino microcontroller boards. Just swap `PC13` for `LED_BUILTIN`.

You could leave the Blue Pill without inserting it in a solderless breadboard because we are not connecting any component or wire to the Blue Pill's ports in the preceding example.

Programming the Curiosity Nano's internal LED

Similar to the Blue Pill, you can use the Curiosity Nano's internal LED to quickly show data from sensors, and so on, without connecting an LED to a port. The whole project containing this example and other supporting files necessary for compiling it on the MPLAB X IDE is stored on the GitHub page. It is a zip file called `16F15376_Curiosity_Nano_LED_Blink_Delay.zip`.

Follow these steps to run the program on the MPLAB X IDE:

1. Connect the USB cable to the Curiosity Nano and insert the board in the solderless breadboard. Unzip the `16F15376_Curiosity_Nano_LED_Blink_Delay.zip` file.

2. On the MPLAB X IDE, click on **File/Open Project** and then open the project.

3. Double-click on the project folder and click on the `Source Files` folder.

4. Click on `main.c` and you will see the following source code:

```c
/*
This program makes the on-board LED to blink once a
second (1000 milliseconds).
Ver. 1. July, 2020. Written by Miguel Garcia-Ruiz
*/
//necessary library generated by MCC:
#include "mcc_generated_files/mcc.h"
void main(void) //main program function
{
    // initializing the microcontroller board:
    SYSTEM_Initialize();
    //it sets up LED0 as output:
    LED0_SetDigitalOutput();
    while (1) //infinite loop
    {
        LED0_SetLow(); //it turns off the on-board LED
        __delay_ms(1000); //it pauses the program for
                          //1 second

        LED0_SetHigh(); //it turns on on-board LED and
                        //RE0 pin

        __delay_ms(1000); //it pauses the program for
                          //1 second
    }
}
```

5. Compile and run the code by clicking on the run icon (colored green), which is on the top menu. If everything went well, you will see Curiosity Nano's internal LED blinking.

As you can see from the preceding example, it has useful C functions specifically created for the Curiosity Nano board, such as the following:

`SetLow(), SetHigh() and __delay_ms().`

Those functions are essential for making projects with microcontroller boards, and they are used in other chapters of this book.

Summary

In this chapter, we learned how to properly configure and set up the MPLAB X and the Arduino IDEs for the C microcontroller board programming. We were introduced to the C programming language, and in particular, a set of C language instructions necessary for programming the Blue Pill and microcontroller boards. To practice what you have learned about the C language, we looked at a number of practical circuits using the boards' internal and external LEDs. The instructions and structure learned in this chapter can be applied to the rest of this book.

Chapter 3, *Turning an LED On and Off Using a Push Button*, will focus on how to connect a push button with a pull-up resistor to a microcontroller board, as well as how to minimize electrical noise when using the push button. It will also explain how to set up a microcontroller board's input port via software, along with possible applications of push buttons.

Further reading

- Gay, W. (2018). *Beginning STM32: Developing with FreeRTOS, libopencm3, and GCC*. St. Catharines, ON: Apress.

- Microchip Technology (2019). *MPLAB X IDE User's Guide.* Retrieved from `https://ww1.microchip.com/downloads/en/DeviceDoc/50002027E.pdf`.

3
Turning an LED On or Off Using a Push Button

In this chapter, we will review and practice how to turn an LED on or off using a **push button** connected to a microcontroller board. A push button is a practical component that acts like a switch and is used for closing or opening an electronic circuit. We can use them to initialize or activate a process in a microcontroller board. Consequently, the input data provided by a push button is important in many microcontroller applications that require human intervention. Specifically, we will cover the following main topics in this chapter:

- Introducing push buttons
- Understanding electrical noise from push buttons
- Connecting an LED to a microcontroller board port and using an internal **pull-up** resistor
- Testing out the push button

By the end of this chapter, you will have learned how to connect a push button to the Curiosity Nano and the Blue Pill microcontroller boards, as well as how to program the push button's input to turn an LED on or off. You will also have learned how to reduce the problem of **electrical noise** in push buttons. Trying to solve this problem is not trivial, as we will see in this chapter. In particular, the *Understanding electrical noise from push buttons* section explains that not all push buttons work 100% free of manufacturing errors, and electrical noise can be present in them when they are used.

Technical requirements

The software tools that we will use in this chapter will be the **MPLAB-X** and **Arduino IDEs**. The code in this chapter can be found in this book's GitHub repository at the following URL:

```
https://github.com/PacktPublishing/DIY-Microcontroller-
Projects-for-Hobbyists/tree/master/Chapter03
```

The Code in Action video for this chapter can be found here: `https://bit.ly/3cXfZLM`

The code examples in this repository will be used to turn an LED on or off using the Curiosity Nano and the Blue Pill microcontroller boards. The IDEs' installation guides and uses were explained in *Chapter 1, Introduction to Microcontrollers and Microcontroller Boards*. In this chapter, we will also use the following pieces of hardware:

- A solderless breadboard.
- The Blue Pill and Curiosity Nano microcontroller boards.
- A micro-USB cable for connecting your microcontroller boards to a computer.
- The ST-Link/V2 electronic interface, which is needed to upload the compiled code to the Blue Pill. Bear in mind that the ST-Link/V2 requires four female-to-female DuPont wires.
- One LED. Any color will do. We prefer to use a red one for our exercises.
- One 220-ohm resistor rated at one-quarter watt.
- Four male-to-male DuPont wires for connecting the resistor and the push button to the boards.
- A regular, **normally open** push button.

The next section provides a brief introduction to push buttons, which are used in electronic circuits.

Introducing push buttons

A push button is an electronic device that basically acts like a mechanical **switch**; it can be used for either closing or opening an electrical or electronic circuit. They are also called *momentary push buttons*, or *pushbuttons*. Push buttons are made with hard materials such as plastic and have a tiny metal spring inside that makes contact with two wires or contacts, allowing electricity to flow through them if the button is pressed (in **normally open** push buttons) or when it is depressed (in **normally closed** push buttons). When the push button is off, the spring retracts, the electrical contact is interrupted, and electrical current will not flow through the contacts. Push buttons are useful for manually controlling or initializing a process in an electrical or electronic circuit, including applications that contain microcontroller boards. The following image shows a normally closed (left) and a normally open (right) push button:

Figure 3.1 – Normally closed (left) and normally open (right) push buttons

As you can see in *Figure 3.1*, the normally open push button (on right) looks depressed. Note that the pins are connected to a microcontroller board.

> **Note**
> Normally open and normally closed push buttons may look exactly the same, depending on their manufacturers and models. If you are unsure, try your button with a microcontroller board and see what type of push button it is. If the push button sends a logic signal to the microcontroller board without you needing to press it, this means that it is a normally closed push button. In this chapter, you will learn how to connect a push button to a microcontroller board.

A typical application of a push button in microcontroller board projects is to connect or disconnect either the ground or a positive voltage from an I/O pin on the microcontroller board. This voltage change that's made by the push button can be seen by the microcontroller board through its I/O pin (port); this initializes a process in the microcontroller board.

There are different types of push buttons in terms of size. Large and robust ones are used for some industrial applications where an operator needs to quickly identify and push them. Smaller buttons are typically used in electrical appliances and devices, such as computer keyboards and landline telephones. In this chapter, we will use a small push button that is commonly found in many electronic kits and in kits that include microcontroller boards. In fact, the Blue Pill and the Curiosity Nano microcontroller boards have small push buttons in their circuits. They can be used in both boards for resetting the programs that run on them. There are two main types of push buttons: **normally open** and **normally closed** push buttons. Let's look at them in more detail:

- **Normally open push buttons**. In this type of button, its switch always remains open when it is not pressed; that is, it does not close an electrical circuit. It makes an electrical contact (closes a circuit) every time we press it. When we press the push button down, its internal switch closes. These are the most common types of push buttons. They can be useful for momentarily activating or initializing a process; for example, pressing the push button briefly to reset a microcontroller board.

- **Normally closed push buttons**. In its default state, this button can close a circuit, meaning that its switch is normally closed without us having to press the push button. We open the switch (hence the circuit where it is connected to) when we press this type of push button. These buttons can be useful when we need to momentarily turn off or interrupt an electrical/electronic circuit. For example, we can open the connection of a sensor to a microcontroller board if we want to stop reading the sensor for whatever reason.

The next section describes a problem that is present in many push buttons, known as **electrical noise**. This type of noise can sometimes be very difficult (but not impossible) to minimize.

Understanding electrical noise from push buttons

Electrical noise can be generated in many push buttons. This can negatively affect the functionality of an electronic circuit where the push button is connected, and it can have unpredictable results in a microcontroller board.

A common problem with push buttons is that *they are not perfect*. They don't close their switch instantly, and in many cases, electrical noise can be produced. This may happen because not all the push buttons are free of manufacturing errors. If we try to connect a push button directly to a microcontroller's I/O port, every time we press the button, it seems that we do it right. To us, it seems like we pressed it only once. However, to the microcontroller board, it looks like the button was pressed many times for extremely short periods of time, and this is because of electrical noise that is generated in the push button. Electrical noise can be defined as random electrical levels or signals coupling with an electronic circuit. This electrical disturbance or interference can vary greatly, from very small voltages to high voltage levels, and its frequency can also change randomly. There are many sources that generate electrical noise, including heat, faulty electronic components, mechanical movement, and loose electrical connections in a circuit, among other sources.

The undesired electrical noise from push buttons is almost always generated by something called **bouncing**, which is caused by the friction and mechanical movements of the push button's internal metal parts and spring. We need to **debounce** the push button to diminish its electrical noise and thus properly close a circuit (if we are using a normally open push button) in a clean and efficient way. If we don't debounce a push button, its internal switch may close a circuit erratically every time we press the push button, which will affect the functionality of the entire circuit or microcontroller board's input. The data signal that's generated by a push button should be either zero volts (logical LOW) or 3.3 volts (logical HIGH). If we don't debounce a push button, it will create electrical noise that possibly changes those logical levels, and thus the microcontroller board may not recognize them as such.

> **Important**
> The HIGH **logic level** voltage used in both the Blue Pill and the Curiosity Nano boards is 3.3 volts. Remember that in some microcontroller boards, their HIGH logic level is 5 volts instead of 3.3 volts, such as the Arduino family of microcontroller boards.

There are several techniques that deal with electrical noise in push buttons. For example, this type of noise can be greatly minimized by either connecting some electronic components to a push button, or via coding, as we will see in the next few sections.

Debouncing a push button via hardware

One way to reduce electrical noise from a push button is to connect a capacitor and two resistors (this is also called **RC debouncing** circuit, a resistor-capacitor network, or an RC filter) connected to the push button, as shown in the following diagram. When we press the push button, the capacitor will be charged. When we release the button, the capacitor will retain its charge for a very short period of time, and the resistor that is connected to it will discharge it after that time. That capacitor's charge represents a HIGH logic voltage, and it can be used in a microcontroller board. Any transient electrical noise that occurs during the capacitor's charge time can be ignored because the capacitor is providing a HIGH logical value in the meantime:

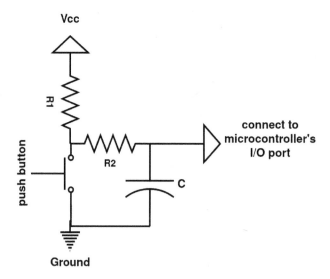

Figure 3.2 – An RC debouncer connected to a push button

The preceding diagram contains two resistors, **R1** and **R2**, a normally open push button, and a capacitor, **C**. The resistors and the capacitor form an RC debouncer. Remember that **Vcc** means positive voltage, which is 3.3 volts for the Curiosity Nano and Blue Pill. Generally, you can obtain Vcc from one of the pins of the microcontroller board labeled 3.3V or Vcc. In addition, you can connect the RC debouncer to one of the microcontroller pins labeled as ground. As we saw in the preceding diagram, the three electronic components can be used to reduce electrical noise in the push button. The typical values for **R1**, **R2**, and **C** are 10K ohms, 10K ohms, and 0.1 microfarads, respectively, although you may need to change those values if the RC debouncer is not working effectively, because the electrical noise is not always the same in each push button. The mathematical formula for calculating the RC debouncer is thoroughly explained Ganssle, J.G. (2008), *A guide to debouncing.*

We included the RC debouncer in this section as a reference, just in case the debouncing method via software does not work for you. The next section will show you how to debounce a push button using software *only*.

Debouncing a push button via software

We can minimize the spurious electrical noise from a push button via coding. A trick that we can use in the code is to ignore the electrical noise for a very short period of time (usually some dozens of milliseconds), right after we press the push button connected to our circuit. The following diagram shows how to connect a push button directly to a microcontroller board's I/O port to perform debouncing via software:

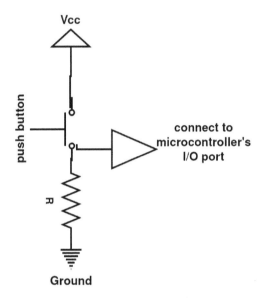

Figure 3.3 – A push button connected to an I/O port with a pull-down resistor

The preceding diagram shows a pull-down resistor, **R**, that is forcing the input port from the microcontroller board to receive zero volts (logical LOW), which is connected to ground, while the push button is not being pressed. A typical value for the pull-down resistor, **R**, is 10k ohms. We can use a pull-down resistor when we need to constantly input a LOW level to the microcontroller board's I/O port, and just change to a logical HIGH level when we press the push button. This can be useful for momentarily starting a process, for example, turning on a light connected to our circuit. The following diagram shows how to connect a pull-up resistor to a microcontroller port, forcing its input to be 3.3 volts (**Vcc**):

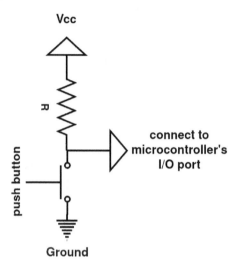

Figure 3.4 – A push button connected to an I/O port with a pull-up resistor

The resistor's value (**R**) in the preceding diagram is typically 10K ohms.

> **Important note**
>
> You need to connect either a pull-down or pull-up resistor to a microcontroller's input port, because if you don't connect anything to the port (this can happen when the push button is not being pressed), the port will present an undetermined (random) voltage at its input due to its internal electronics arrangement. This is called a **floating** input voltage.

The preceding diagram includes a normally open push button. Once the push button is pressed, the input voltage will change to a logical HIGH level, or 3.3 volts. Remember that the logical HIGH level can be 5 volts, depending on the microcontroller board that you are using. The **Vcc** voltage and the ground are connected to the microcontroller board.

Fortunately, many microcontrollers provide internal pull-up and pull-down resistors that are connected to their I/O ports that can be activated via coding. The Blue Pill contains both! This means that we can connect a push button directly to its I/O ports, without connecting an external resistor, and just activate the board's internal pull-up or pull-down resistors via software. This speeds up prototyping development that requires the use of push buttons, but this may not always be the ideal solution.

The following diagram shows two ways of connecting a push button directly to an I/O port. This port already has a pull-up or a pull-down resistor that's been activated via software:

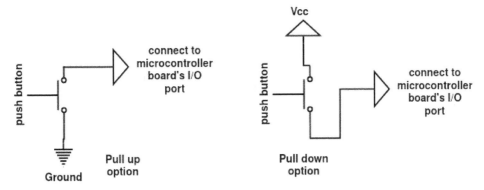

Figure 3.5 – Push buttons connected directly to an I/O port

As you can see, the debouncing method via software does not use the RC debouncing that's connected to the push button. This method works most of the time, and it saves you time and effort. However, you should experiment and try out both methods if the electrical noise from the push button is persistent.

Each debouncing method has its trade-offs. In the hardware-based method, you will need to connect some electronic components to the push button. You will also need to *buy* those extra components. In the software-based method, where you will be using the microcontroller's internal pull-up or pull-down resistors, you don't need to connect any extra components to the push button, but you will need to add more lines of code to your program to deal with the debouncing, and these instructions may take up some valuable processing cycles from your microcontroller. Despite this, we recommend that you use the software-based debouncing method because it is simple to implement.

The next section deals with examples that show you how to connect a push button to the Blue Pill and Curiosity Nano microcontroller boards, as well as how to debounce it via software.

Connecting an LED to a microcontroller board port and using an internal pull-up resistor

In this section, you will learn how to connect a push button to both the Blue Pill and the Curiosity Nano boards. This is a simple exercise for these microcontroller boards, and it demonstrates how to use a push button to send a logical LOW level signal to a microcontroller board to turn an LED connected to it on or off. If we want to use a push button in our electronic circuit example, we will need to connect it to an input port from a microcontroller board. Also, remember that we should debounce our push button to avoid undesirable results.

The following subsection will show you how to debounce a push button that is connected to the Blue Pill via coding. This is the simplest way to debounce a push button and you can use this method in other chapters of this book.

Debouncing a push button via software connected to the Blue Pill

In this section, we will show you a Fritzing diagram, and then a photo that shows how everything is connected. We will also look at some code that demonstrates how to debounce the push button via software.

The following is a Fritzing diagram that shows how to connect a push button directly to an I/O port that is already using its internal pull-up resistor. The LED and its respective resistor are connected to the Blue Pill's port number; that is, B12:

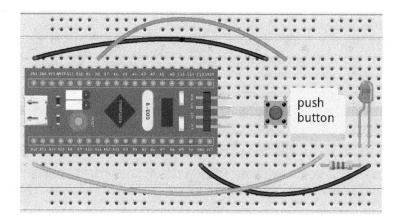

Figure 3.6 – A Blue Pill using its internal pull-up resistor with the push button

As you can see, the LED will turn on or off every time we press the push button.

Please note that in the preceding image, the push button is connected to the center of the breadboard, to the right of the microcontroller board. Follow these steps to connect the push button and the LED to the Blue Pill, as shown in the preceding image:

1. Insert the Blue Pill into the solderless breadboard.

2. Insert the push button into the breadboard and connect one of its pins to the Blue Pill's ground pin using a DuPont wire.

3. Connect another pin from the push button to port B0 of the Blue Pill using a DuPont wire.

4. Insert a 220-ohm resistor into the breadboard and connect one of its pins to port B12 of the Blue Pill using a DuPont wire.

5. Insert the LED into the breadboard, connecting its anode to the other resistor's pin.

6. Connect the LED's cathode to one of the ground pins of the Blue Pill using a DuPont wire.

The following image shows how to connect the push button to the Blue Pill, based on the Fritzing diagram shown previously:

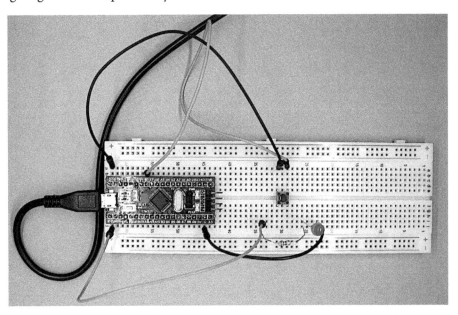

Figure 3.7 – Connecting the push button to the Blue Pill's cathode, and then to one of the ground pins from the Blue Pill using a DuPont wire

As you can see, the Blue Pill board has ground (G) pins on the top and bottom rows of pins. This facilitates the component connections in the circuit.

Remember that you will need to connect the ST-Link/V2 electronic interface to the Blue Pill to upload the program to it from the Arduino IDE, as explained in *Chapter 1, Introduction to Microcontrollers and Microcontroller Boards*.

The following code shows how to debounce a push button via software on the Blue Pill. You can find this code file in this book's GitHub repository, with comments provided. The file is called `internal_pullup_debounce_Blue_Pill.ino`:

```
#define PinLED PB12
#define Pinpushbutton PB0


void setup() {
  pinMode(PinLED, OUTPUT);
  pinMode(Pinpushbutton, INPUT_PULLUP);
}


int reading_pushbutton;
int ledState = HIGH;
int buttonState;
int lastButtonState = LOW;
unsigned long lastDebouncingTime = 0;
unsigned long debouncingDelay = 50;


void loop() {

reading_pushbutton=digitalRead(Pinpushbutton);
if (reading_pushbutton!= lastButtonState) {
    lastDebouncingTime = millis();
}
if ((millis() - lastDebouncingTime) > debouncingDelay) {
    if (reading_pushbutton!=buttonState) {
        buttonState = reading_pushbutton;
        if (buttonState == HIGH) {
            ledState = !ledState;
        }
```

```
        }
    }
  digitalWrite(PinLED, ledState);
  lastButtonState = reading_pushbutton;
}
```

The preceding code waits 50 milliseconds once the push button has been pressed and toggles the LED value. This value is experimental, so you may need to change it if your push button is working erratically.

> **Important Note**
>
> The Blue Pill's I/O ports are referenced in the Arduino IDE with the letter P. For example, port B12 is referenced as PB12 in the IDE. In addition, the port labels (names) must be written in capital letters.

As shown in the preceding code, it continuously reads the port B0 of the Blue Pill. If the push button is pressed, port B0 is connected to ground by the push button. Then, the B12 output port sends out a HIGH logical level and turns on the LED connected to B12. If the push button is not pressed, the B12 port sends out a LOW logical level.

> **Tip**
>
> You can also debounce a push button via software if you are using an Arduino microcontroller board. In fact, the software-based debouncing method we used in this chapter is based on the method that's used in Arduino boards, as explained here:
>
> https://www.arduino.cc/en/Tutorial/
> BuiltInExamples/Debounce

If your LED turns on and off when you press the push button, congratulations! Pay attention to how the microcontroller board reacts when you press the push button. If the LED turns on erratically several times when you just press the push button, you may need to change either the resistor or capacitor values if you are doing RC debouncing, or change the millisecond waiting value if you are debouncing the push button via software.

The next section describes how to debounce a push button via software connected to the Curiosity Nano microcontroller board.

Turning an LED on or off with a push button on the Curiosity Nano

In this section, we use the Curiosity Nano to debounce a push button via software by waiting some milliseconds once the push button has been pressed. We can use the __delay_ms() function for this. Remember that the function is written with two underscore symbols (__).

The following Fritzing diagram shows how to connect the push button to the Curiosity Nano:

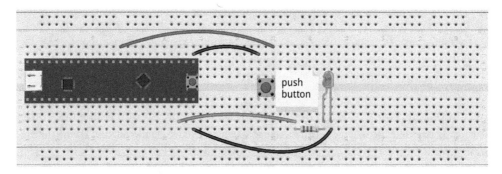

fritzing

Figure 3.8 – A push button directly connected to the Curiosity Nano board

Please note that the push button is connected to the center of the breadboard, to the right of the microcontroller board.

According to the preceding image, here are the steps for connecting all the components:

1. Insert the Curiosity Nano into the solderless breadboard.
2. Insert a push button into the breadboard and connect one of its pins to the Curiosity Nano's ground pin using a DuPont wire.
3. Connect another pin from the push button to port RA0 of the Curiosity Nano using a DuPont wire.
4. Insert a 220-ohm resistor into the breadboard and connect one of its pins to the port RD2 of the Curiosity Nano using a DuPont wire.
5. Insert the LED into the breadboard, connecting its anode to the other resistor's pin.
6. Connect the LED's cathode to one of the ground pins of the Curiosity Nano using a DuPont wire.

The following image shows how everything is connected:

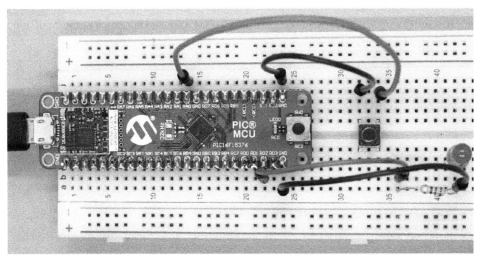

Figure 3.9 – The Curiosity Nano and a push button connected to it

In the preceding image, you can see that the Curiosity Nano has ground (GND) pins in both the upper and lower rows of pins. This allows us to wire the components in the circuit.

We have created a project for the MPLAB-X IDE, which can found in this book's GitHub repository. It contains comments explaining each line of code. You will need to uncompress its ZIP file first to open the project in the MPLAB-X IDE. This project file is called 16F15376_Curiosity_Nano_pushbutton.X.zip.

The following code from that project shows how software-based debouncing is done:

```
#include <xc.h>
#include <stdio.h>
#include "mcc_generated_files/mcc.h"
int reading_pushbutton=0;
void main(void)
{
    SYSTEM_Initialize();
    IO_RD2_SetDigitalOutput();
    IO_RA0_SetDigitalInput();
    IO_RA0_SetPullup();
    IO_RD2_SetLow();
    while (1)
    {
```

```
reading_pushbutton=IO_RA0_GetValue();
__delay_ms(100);
reading_pushbutton=IO_RA0_GetValue();
if (reading_pushbutton==LOW){
    IO_RD2_Toggle();
}
}
}
```

The preceding code reads the value from the push button and waits 100 milliseconds; then, it reads it again to see if the push button is still being pressed. We found this 100-millisecond value experimentally and it seems to work most of the time. Remember that you may need to adjust it, depending on how your own push button behaves in your circuit.

This is a slightly different approach than the one we used in the Blue Pill. We coded the waiting time to try and ignore some electrical noise that may occur during that time. If your LED turns on and off when you press the push button, congratulations! You are now able to connect and use a push button in an electronic circuit connected to a microcontroller board. Remember that a push button may be used to initiate a process or activity in a microcontroller board.

The next section will show you how to check if a push button is working OK, and if it is a normally open or normally closed push button.

Testing out the push button

In this section, we will focus on testing a push button. Before using it with a microcontroller board, it's a good idea to try it out to see if it works mechanically, and testing allows us to find out if the push button is normally closed or normally opened. The following image shows how to connect all the components to try out the push button:

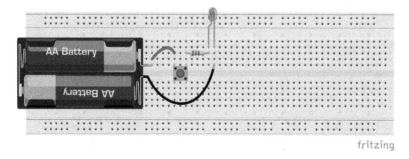

Figure 3.10 – Connecting the push button to an LED and a battery pack

As you can see, we don't need to connect a push button to a microcontroller board to test it. Here are the steps for connecting the components and testing the push button:

1. Connect the batteries' positive (+) terminal to one pin of the push button.

2. Connect the other push button pin to the 220-ohm resistor.

3. Connect the 220-ohm resistor to the LED's anode pin.

4. Connect the LED's cathode pin to the batteries' negative (-) terminal. Be careful when connecting the LED's pins. If it is connected in reverse, the LED will not turn on.

5. Once you've connected everything, if the LED turns on without you pressing the push button, this means that the push button is of the normally closed type. If it does, the LED should turn off when you press the push button. If the LED turns on every time you press the push button, this means that it is a normally open one.

6. Press the push button several times. If the LED turns on and off erratically, or if the LED does not turn on at all, the push button may be faulty and you will need to replace it, assuming that the batteries have enough voltage.

Connecting the batteries to the push button and the LED should be enough to see if the push button works.

Summary

In this chapter, we learned what a push button is and how can we reduce the problem of electrical noise that many push buttons have, a process called debouncing. This process can be done either via software or hardware. We also reviewed the importance of push buttons in some electronic projects that require human intervention – for example, how to manually restart a microcontroller board when we press its on-board push button. Using a push button in an electronic circuit containing a microcontroller board is important because you are programming a user interaction there, so a user can start a process in the microcontroller board.

The next chapter will focus on how to connect a photoresistor (a sensor that measures the amount of light in the environment).

Further reading

- Ganssle, J.G. (2008). *A guide to debouncing.* Technical Report. Baltimore, MD: The Ganssle Group.

- Gay, W. (2018). *Beginning STM32: Developing with FreeRTOS, libopencm3 and GCC.* St. Catharines: Apress.

- Horowitz, P., Hill, W. (2015). *The art of electronics. [3rd ed.]* Cambridge University Press: New York, NY.

- Microchip (2019). *PIC16F15376 Curiosity Nano Hardware User Guide.* Microchip Technology, Inc. Available from `http://ww1.microchip.com/downloads/en/DeviceDoc/50002900B.pdf`

- Mims, F.M. (2000). *Getting started in electronics.* Lincolnwood, IL: Master Publishing, Inc.

- Ostapiuk, R. & Tay, I. (2020). *Fundamentals of the C programming language.* Microchip Technology, Inc. Retrieved from `https://microchipdeveloper.com/tls2101:start`

- Ward, H.H. (2020). *C programming for the PIC microcontroller.* New York, NY: Apress.

4
Measuring the Amount of Light with a Photoresistor

This chapter focuses on how to connect a **photoresistor**, an electronic component that measures the amount of light from the environment, to an input port of both the Blue Pill and Curiosity Nano microcontroller boards. In this chapter's exercise, we will analyze with a photoresistor whether a plant receives enough light.

In this chapter, we are going to cover the following main topics:

- Understanding sensors
- Introducing photoresistors
- Connecting a photoresistor to a microcontroller board port
- Coding the photoresistor values and setting up ports
- Testing the photoresistor

By the end of this chapter, you will have learned how to connect an analog sensor to a microcontroller board, and how to analyze analog data obtained from a photoresistor. The knowledge and experience learned in this chapter will be useful in other chapters that require the use of a sensor.

Technical requirements

The software tools that you will be using in this chapter are the Arduino IDE and the MPLAB-X for editing and uploading your programs to the Blue Pill and the Curiosity Nano boards, respectively.

The code used in this chapter can be found in the book's GitHub repository here:

```
https://github.com/PacktPublishing/DIY-Microcontroller-
Projects-for-Hobbyists/tree/master/Chapter04
```

The Code in Action video for this chapter can be found here: `https://bit.ly/3gNY4bt`

In this chapter, we will also use the following pieces of hardware:

- A solderless breadboard.

- The Blue Pill and Curiosity Nano microcontroller boards.

- A micro-USB cable for connecting your microcontroller boards to a computer.

- The ST-Link/V2 electronic interface needed for uploading the compiled code to the Blue Pill. Bear in mind that the ST-Link/V2 requires four female-to-female DuPont wires.

- A green, yellow, and red LED.

- Three 220-ohm resistors rated at one-quarter watt. These resistors are for the 3 LEDs.

- One 220-ohm resistor rated at one-quarter watt for the photoresistor connected to the Curiosity Nano.

- One 10 K ohm resistor, rated at one-quarter watt.

- A **photoresistor sensor module**.

- Three male-to-female DuPont wires for connecting the sensor module to the solderless breadboard.

- A dozen male-to-male DuPont wires for connecting the resistor and the push button to the breadboard and the microcontroller boards.

The next section describes a brief introduction to photoresistors, their electrical characteristics, and how they are used in electronic projects.

Understanding sensors

In this section, we explain what sensors are and the importance of sensor modules. It is important to understand first what sensors are and what their purpose is before you use them in practical microcontroller board applications, for example, using photoresistors, which are a useful type of sensor. This information about sensors is essential for further sections.

Defining sensors

A **sensor** is an electronic component, device, or module that measures physical input from an environment or a particular situation (for example, a sensor measuring water temperature in a fish tank). Sensors are useful for detecting changes in physical variables, such as humidity, temperature, light, vibrations, and movement, among others. Those physical variations are manifested in the sensors by changing their electric/electronic properties, such as changes in resistance and conductivity in the sensors.

There are different types of sensors with different applications. For example, motion sensors can detect human or pet movements when they pass across the sensor's field of view. If this sensor detects motion, it sends a signal to a computer or microcontroller board and then the board should do something about it, such as open an automatic door, set off an alarm, turn on a lightbulb, and so on. Other types of sensors include infrared, ultrasonic, temperature, pressure, and touch sensors.

What are sensor modules?

A sensor can be part of a small electronic circuit called a **sensor module**. It contains other electronic components besides a sensor, such as resistors, transistors, LEDs, and integrated circuits, among others. The purpose of those extra components in a sensor module is to support and facilitate the reading, analysis, and transmission of signals coming from a sensor. Some sensor modules convert analog to digital data. Analog data are voltage variations, that is, analog to physical variables from an environment. For example, 0.5 volts obtained from a temperature sensor could be equivalent to 25 degrees Celsius. Digital data coming from a sensor module can contain either logical-level *HIGH* or logical-level *LOW* values.

> **Note**
>
> Voltages representing digital logic levels change depending on the microcontroller board that you are using. The I/O ports from both the Blue Pill and Curiosity Nano microcontroller boards use 3.3 volts, which is equivalent to logical-level HIGH. Logical-level LOW is equivalent to 0 volts.

An analog signal from a sensor is sent electronically via a wire or a wireless communication medium (for example, Bluetooth) to a microcontroller board, which will process it, and then do something about it, as shown in *Figure 4.1*:

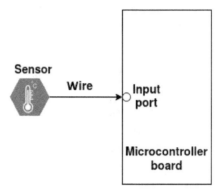

Figure 4.1 – A temperature sensor connected to the microcontroller board's port

The data coming from the sensor (displayed in the diagram from *Figure 4.1*) is read by the microcontroller board's input port. This data can be adapted and shown to be human-readable (for example, showing the temperature numerically in a display as degrees Celsius or Fahrenheit). The same happens with analog and/or digital signals coming from a sensor module. Its data is sent to a microcontroller board port or ports via a wired or wireless medium.

> **Note**
> Microcontroller boards' ports can be set up as either input or output ports via coding. Please refer to *Chapter 2, Software Setup and C Programming for Microcontroller Boards*, on how to program them.

The next section focuses on photoresistors, a commonly used type of sensor, describing their function, representation, and applications.

Introducing photoresistors

This section introduces you to photoresistors, which are very useful for many applications, for example, for measuring the amount of light. In this chapter, we will define what photoresistors are, their classification, and how they are connected to an electronic circuit.

A **photoresistor** is an electronic component made with light-sensitive material, changing its resistance according to the amount of visible light that it detects. There are different types of photoresistors. Some of them detect **ultraviolet** (**UV**) light and others detect infrared light. The latter is used in TV sets, where its infrared light sensor receives data from a remote control. *Figure 4.2* shows a common photoresistor used in the examples of this chapter. The photoresistor used in this chapter detects regular light that humans can see. It does not detect infrared nor UV light.

Figure 4.2 – A typical photoresistor

From *Figure 4.2*, you can see that the photoresistor has two pins (also called legs). They are connected to an electronic circuit similar to regular resistors, so photoresistors do not have polarity. They work as a regular resistor that changes its resistance according to the amount of light it receives through its transparent cover. Here are some technical specifications of the photoresistor:

- Size: 5mm diameter (width)

- Range of resistance: 200 K ohms (dark) to typically 5 K to 10 K ohms and nearly 0 ohms when it receives full brightness.

- Power supply: Up to 100V, using less than 1 milliamp of current on average, depending on the power supply voltage.

Figure 4.3 shows an electrical diagram containing a photoresistor and how it can be connected to a microcontroller board.

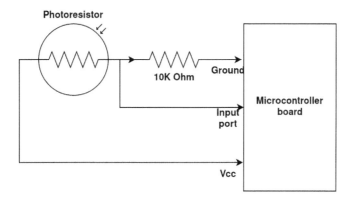

Figure 4.3 – Electrical diagram with a photoresistor

As you can see in *Figure 4.3*, a photoresistor's electrical symbol is represented with a circle and some incoming arrows indicating that light rays are hitting the photoresistor's surface. The photoresistor is connected to the power supply to a **Ground** pin from a microcontroller board and the other pin is connected to a voltage pin (**Vcc**) and an **Input port** from a microcontroller board. That is one way to connect a photoresistor. The **10K ohm** resistor works as a pull-down resistor. This is because the input port where the photoresistor is connected to at some point will receive zero volts when the photoresistor is not receiving any light. Remember that we should not leave an input pin without connecting to anything, otherwise its state will be floating (inputting random voltages).

> **Note**
>
> The photoresistor should be connected to an analog input port from the microcontroller board since the port will receive a changing analog voltage from the photoresistor because the resistance presented at the photoresistor will change according to the amount of light received by the photoresistor.

The next section explains how to connect and use photoresistors in both the Blue Pill and Curiosity Nano boards to measure the amount of light in an environment.

Connecting a photoresistor to a microcontroller board port

This section shows how to connect a photoresistor to the Blue Pill and the Curiosity Nano boards to *read* the amount of light from an environment (for example, a living room).

In this section, we also explain how to use three LEDs to indicate that a room is well illuminated by turning a green LED on, that the light is dim by turning a yellow LED on, or that it is dark by turning a red LED on. The electronic circuit with the photoresistor sensor can be placed close to a plant and the circuit can be useful to know whether a plant needs more light or not. The next section shows how to build the circuit with the Blue Pill board.

Connecting a photoresistor to a Blue Pill board

The connection of a photoresistor sensor to a Blue Pill microcontroller board is simple. It can be directly connected to an input analog port, provided that you use a pull-down resistor. The circuit shown in *Figure 4.4* describes how to do it, and it is based on the electrical diagram from *Figure 4.3*.

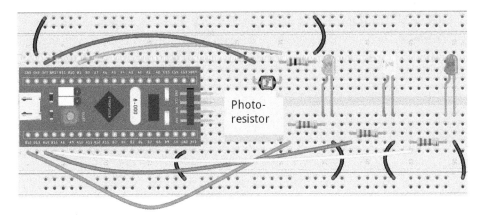

Figure 4.4 – The LEDs and photoresistor connected to a Blue Pill board

As you can see in *Figure 4.4*, the three LEDs will show the amount of light detected by the photoresistor. The following are the steps for connecting the components to the Blue Pill board:

1. Connect the Curiosity Nano's GND pins to the upper and lower rails, of the solderless breadboard, as shown in *Figure 4.4*.

2. Connect one pin of the photoresistor to the microcontroller board's ground pin labeled **3V3**.

3. Connect the other pin of the photoresistor to the Blue Pill's pin labeled **B1**. **B1** will be used as an analog input port.

4. Connect a **10K ohm** resistor to the ground and to the photoresistor's leg that is connected to input port **B1** of the Blue Pill.

5. Now, connect the resistors to the output ports **B12**, **B14**, and **B15** of the Blue Pill.

6. As the last step, connect the green, yellow, and red LEDs' anodes to the three resistors and then connect the LEDs' cathodes to the ground.

The 3V3 pin from the Blue Pill provides 3.3 volts, which are enough for applying voltage (feeding) to the photoresistor.

> **Important note**
>
> Do not apply **5 volts (5V)** to the Blue Pill's input ports, because you may damage the Blue Pill. That's why you will connect the 3V3 voltage pin to the photoresistor, so it will provide up to 3.3 volts as an analog voltage output and the Blue Pill will use up that voltage to measure the amount of light. Remember that the voltage coming from the photoresistor will change according to its resistance.

We are connecting a **10K ohm** pull-down resistor to the Blue Pill's input port **B1**, as shown in *Figure 4.4*, forcing the port to have a 0-volt input when there's no voltage coming from the photoresistor. Remember that the input ports should receive some voltage or even 0 volts to avoid random voltages.

> **Note**
>
> You may want to try a different value for the pull-down 10K ohm resistor connected to the Blue Pill, shown in *Figure 4.4*, depending on the light level range that you would like to detect.

Figure 4.5 shows how all the components are connected to the microcontroller board. This circuit is based on the Fritzing diagram shown in *Figure 4.4*.

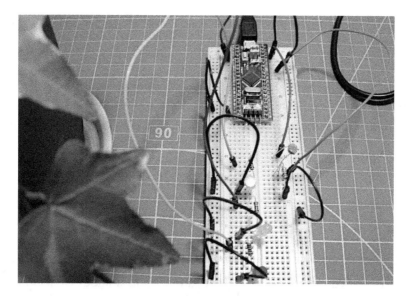

Figure 4.5 – Connecting the photoresistor and the LEDs to the Blue Pill

As you can see in *Figure 4.5*, the photoresistor is connected to the Blue Pill's port B1. The green LED is turning on, meaning that the amount of light that the plant is receiving should be OK. Remember that you will need to connect the ST-Link/V2 electronic interface to the Blue Pill in order to upload the program to it from the Arduino IDE, as explained in *Chapter 1, Introduction to Microcontrollers and Microcontroller Boards*.

In this section, you learned how to connect an easy-to-use photoresistor to the Blue Pill board and how to show its values through LEDs.

The next section describes how to code the photoresistor example shown in *Figure 4.4* and *Figure 4.5*.

Coding the photoresistor values and setting up ports

This section shows how to code a Blue Pill application for reading data from a photoresistor.

The following code should run on the microcontroller board circuit shown in *Figure 4.4* and *Figure 4.5*:

```
int photoresistorPin = PB1;
int photoresistorReading;
void setup(void) {
    pinMode(PB12, OUTPUT);
    pinMode(PB14, OUTPUT);
    pinMode(PB15, OUTPUT);
}
void loop(void) {
    photoresistorReading = analogRead(photoresistorPin);
    digitalWrite(PB12, LOW);
    digitalWrite(PB14, LOW);
    digitalWrite(PB15, LOW);
    if (photoresistorReading < 600) {
        digitalWrite(PB15, HIGH);
    } else if (photoresistorReading < 1000) {
        digitalWrite(PB14, HIGH);
    } else {
        digitalWrite(PB12, HIGH);
    }
    delay(500);
}
```

As you can see from the code, one of its most important functions is this one: `analogRead(photoresistorPin);`. Internally, that function will perform an **analog-to-digital conversion (ADC)** of the voltage coming from the photoresistor. That ADC will map input voltages between 0 and 3.3 volts into integers between 0 and 4095. We defined these thresholds experimentally: if the ADC value is < 600, it means that the environment is dark. If it is between 600 and 1000, it means that the environment is dim. If the value is > 1000, it means that the environment is well illuminated. You may need to change those thresholds if your environment lighting changes, so you will need to experiment a bit with those values. We use the `pinMode()` function to declare the PB12, PB14, and PB15 ports as outputs.

The code can be downloaded from the book's GitHub repository. Its filename is `photoresistor_bluepill.ino`. The code from the repository contains useful comments explaining the functions and variables used in it.

Note

You can run the same code for the Blue Pill on Arduino microcontroller boards such as the Arduino Uno. Just change the Blue Pill's input port PB1 for the Arduino's analog port 0 (it is labeled as A0 on the Arduino Uno), and change the Blue Pill's output ports PB12, PB14, and PB15 for the Arduino digital ports 8, 9, and 10. You may also need to change the values (thresholds) for the decisions from the code.

The following section describes how to connect the LEDs and a photoresistor to the Curiosity Nano microcontroller board, following the example from the Blue Pill board.

Connecting a photoresistor to a Curiosity Nano board

Following the Blue Pill example from the previous section, we can show the amount of light in an environment by turning on LED lights using the Curiosity Nano microcontroller board, as depicted in *Figure 4.6*.

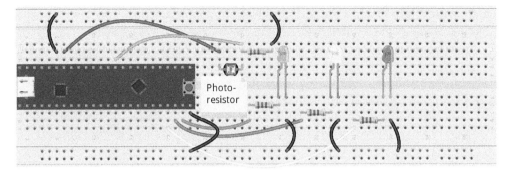

fritzing

Figure 4.6 – The LEDs and photoresistor connected to a Blue Pill board

As it is shown in *Figure 4.6*, the three LEDs will be used to show the amount of light detected by the photoresistor. The following are the steps for connecting the components to the Curiosity Nano board:

1. Connect the Curiosity Nano's GND pins to the upper and lower rails of the solderless breadboard, as shown in *Figure 4.6*.

2. Connect one pin of the photoresistor to the Curiosity Nano's pin labeled VTG. This pin provides 3.3 volts.

3. Connect the other pin of the photoresistor to the Curiosity Nano's pin labeled RA0. RA0 will be used as an analog input port.

4. Connect a 220-ohm resistor to the ground and to the photoresistor's leg that is connected to port RA0. This will be a pull-down resistor.

5. Now, connect the 220-ohm resistors to ports RD1, RD2, and RD3 and to the anodes of the 3 LEDs, respectively.

6. As the last step, connect the LEDs' cathodes to the ground.

We are connecting a 220-ohm pull-down resistor to the Curiosity Nano's input port, as shown in *Figure 4.6*, forcing the port to have a 0-volt input when there's no voltage coming from the photoresistor. Remember that the input ports should receive some voltage or even 0 volts to avoid random voltages.

> **Note**
> You may want to try a different value for the pull-down 220-ohm resistor shown in *Figure 4.6*, depending on the light level range that you would like to detect.

The photoresistor from *Figure 4.6* will change its resistance according to the amount of light that it receives from the environment, thus the voltage that passes through it will change. These voltage differences will be read by a microcontroller board's input port. Our code running on the microcontroller board will then compare the analog voltage values and determine whether the light is very low, normal, or too bright by turning on the yellow, green, or red LEDs respectively.

The next code shows how to read the photoresistor values through the analog port RA0:

```c
#include <xc.h>
#include <stdio.h>
#include "mcc_generated_files/mcc.h"
static uint16_t reading_photoresistor=0;
void main(void)
{
    SYSTEM_Initialize();
    ADC_Initialize();
    while (1)
    {
        IO_RD1_SetLow();
        IO_RD2_SetLow();
        IO_RD3_SetLow();
        reading_photoresistor =
            ADC_GetConversion(channel_ANA0);

        if (reading_photoresistor>=0 &&
                reading_photoresistor <=128)
        {
            IO_RD1_SetHigh();
        } else if (reading_photoresistor>= 129 &&
                reading_photoresistor<=512)
        {
            IO_RD2_SetHigh();
        } else
        {
            IO_RD3_SetHigh();
        }
```

```
            __delay_ms(200);
    }
}
```

Please bear in mind that the code uploaded to the book's GitHub online repository has many comments explaining almost all its instructions. In the preceding code, one of the most important functions is `ADC_GetConversion(channel_ANA0);`, which reads the voltage changes from the photoresistor and performs the analog to digital conversion of those values. `channel_ana0` is a label given to port RA0.

The code can be downloaded from the book's GitHub repository. Its filename is `Chapter4_Curiosity_Nano_code_project.zip`. The code from the repository contains useful comments explaining the functions and variables used in it.

Figure 4.7 shows how the photoresistor and the LEDs are connected to the Curiosity Nano's ports. The circuit shown in *Figure 4.7* is based on the Fritzing diagram shown in *Figure 4.6*.

Figure 4.7 – Connecting the photoresistor and the LEDs to the Curiosity Nano

As you can see in *Figure 4.7*, the photoresistor is connected to the Curiosity Nano's port RA0. The green LED is turning on in the figure, meaning that the amount of light that the plant is receiving should be OK.

The next section shows how to obtain analog data about the amount of light from an environment using a photoresistor sensor module.

Connecting a photoresistor sensor module to the microcontroller boards

This section explains how to use a photoresistor sensor module for measuring the amount of light from an environment. This module can be bought separately or as part of a sensors kit from many online sources. The photoresistor sensor module contains tiny electronic components that facilitate the connection to a microcontroller board and its photoresistor use. For example, the module that we are using in this section contains extra components such as resistors and a variable resistor that adjusts the threshold level of a digital *high* value, which is sent by the sensor module if it receives a certain amount of light. Of course, this digital value must be sent to a digital input port of the microcontroller board.

The next section shows how to connect the photoresistor sensor module to the Blue Pill.

Connecting the photoresistor sensor module to the Blue Pill board

This section explains how to connect a photoresistor module to the microcontroller board using just three wires. The photoresistor sensor module shown in *Figure 4.8* has four connectors: **GND** (short for **ground**), **A0** (short for **analog output**), **D0** (short for **digital output**), and **+5V**. The order of these connectors may change if your sensor module is from a different brand, but their purpose is the same.

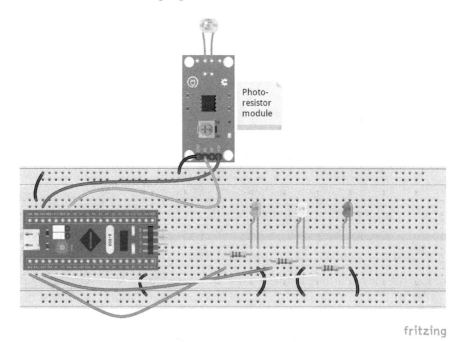

Figure 4.8 – A photoresistor sensor module connected to a Blue Pill board

Here are the steps on how to connect the sensor module to the Blue Pill:

1. Connect the sensor module's GND pin (in some sensor modules, it is labeled with the – symbol) to the microcontroller board's ground connector labeled GND.

2. Connect the sensor module's +5V (in some modules, it is labeled with the + symbol) pin to the 3V3 voltage pin from the microcontroller board. Even though the 3V3 pin provides 3.3 volts, it is enough for applying voltage (feeding) to the sensor module.

> **Important note**
> Do not apply **5 volts (5V)** to the Blue Pill's input ports, because you may damage the Blue Pill. That's why you will connect the 3V3 voltage pin to the sensor module's +5V pin, so the module will provide up to 3.3 Volts as an analog voltage output, and the Blue Pill will use up that voltage to measure the amount of light.

3. Connect the module's A0 pin to the Blue Pill's B1 pin (which is an input port).

4. Now, connect the resistors to Blue Pill ports B12, B14, and B15. They are output ports.

5. As the last step, connect the green, yellow, and red LEDs' anodes to those resistors and then connect the LEDs' cathodes to the ground.

As you can see in *Figure 4.8*, we are not connecting a pull-down resistor to the Blue Pill's input port because the sensor module already contains a pull-down resistor. This allows us to save some space in the circuit and time to connect extra components to it.

The following section describes how to use the sensor module with the Curiosity Nano.

Connecting the photoresistor sensor module to the Curiosity Nano board

In this section, we will analyze how the sensor module's analog pin is connected to the Curiosity Nano, as is shown in *Figure 4.9*. Please remember that this circuit does not need the pull-down resistor. We are using the same input port that was used for connecting the single photoresistor, described in an earlier section of this chapter. *Figure 4.9* shows a Fritzing diagram that includes the photoresistor module connected to the Curiosity Nano board.

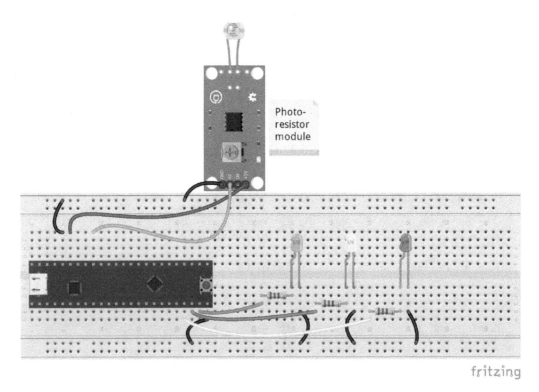

Figure 4.9 – The photoresistor module connected to Curiosity Nano's RA5 input port

Following the Fritzing diagram from *Figure 4.9*, these are the steps for connecting the photoresistor sensor module to the Curiosity Nano:

1. Connect the module's GND pin (in some modules, it is labeled with the - symbol) to the microcontroller board's ground connector, labeled GND.

2. Connect the module's +5V pin (in some modules, it is labeled with the + symbol) to the VTG pin from the microcontroller board. Even though the VTG pin provides 3.3 volts, it may be enough to apply voltage (feeding) to the sensor module.

> **Important note**
>
> Do not apply 5V to the Curiosity Nano's input ports because you may damage the microcontroller board. That's why you will connect the Curiosity Nano's VTG voltage pin to the sensor module's +5V pin, so the sensor module will provide up to 3.3 Volts as an output and the Curiosity Nano will use up that voltage to measure the amount of light.

3. Connect the sensor module's A0 pin to the Curiosity Nano's RA5 pin (which is an input port).

4. Now, connect the resistors that will protect the 3 LEDs to ports RD1, RD2, and RD3. They are output ports.

5. As the last step, connect the green, yellow, and red LEDs' anodes to those resistors and then connect the LEDs' cathodes to the ground.

As you can see from *Figure 4.9*, similar to the example with the Blue Pill, we are not connecting a pull-down resistor to the Curiosity Nano's input port because the sensor module already contains a pull-down resistor.

You can use the same code used for the photoresistor used in *Figure 4.6* and *Figure 4.7* to get analog values from the photoresistor sensor module shown in *Figure 4.9*. You may need to adjust the values in the code if necessary. You can experiment with a number of environments (for example, using the sensor in a living room or a bedroom) to see the analog values obtained by the microcontroller board from the sensor change and thus adjust the code accordingly.

The next section describes how to test the photoresistor if something is wrong with it, for example, the LEDs are not turning on, as a way to troubleshoot it.

In this section, you learned how to connect the photoresistor module and the LEDs to the Curiosity Nano's ports. This section is important because it shows how practical and easy to use the photoresistor modules are, which facilitate the connections in microcontroller board projects.

Testing out the photoresistor

This section focuses on how to test out a photoresistor to see if it is working OK. First of all, remember that the photoresistor used in this chapter does not have polarity, so you can safely connect any of its pins (legs) to a microcontroller board's input port.

You also need to make sure that the pull-down resistor connected to the photoresistor has the right value. For example, the pull-down resistor used in the Blue Pill example from *Figure 4.4* is 10K ohm, and we used a 220-ohm resistor for the Curiosity Nano example from *Figure 4.6*. We found those resistor values experimentally. You can try out different resistors connected to the photoresistor to see if the voltage passing through the photoresistor changes widely. Ideally, that voltage should be changing between 0 and 3.3 volts, or close to those values, because in our circuit examples from this chapter, we connected one pin of the photoresistor to 3.3 volts.

In order to see if the photoresistor is working OK, you can use a multimeter. Follow the next steps for testing the photoresistor with a multimeter:

1. Connect the multimeter's red probe (test lead) to one pin of the photoresistor.
2. Connect the multimeter's black probe to the other pin of the photoresistor.
3. Turn on the multimeter and set it up for measuring resistance (ohms).
4. Cover the photoresistor with your hand and uncover it. Its resistance should change in the multimeter because the light the photoresistor is receiving changes.

If you don't have a multimeter, you can test out a photoresistor with just a voltage source, such as batteries, and an LED (any color will do). *Figure 4.10* shows how to connect the photoresistor to the LED:

Figure 4.10 – Testing out the photoresistor

As you can see from *Figure 4.10*, the light from the LED should change (for example, changing from dim to bright) when the photoresistor receives different amounts of light, so try to cover it with your hand and see what happens. If the LED light does not change, it is probable that the photoresistor is damaged, so you will need to replace it. Here are the steps for connecting the components:

1. Connect the positive (+) battery terminal to one pin of the photoresistor.

2. Connect the other photoresistor pin to the LED's anode pin.

3. Connect the LED's cathode pin to the negative (-) battery terminal.

Be careful when connecting the LED's pins. If it is connected in reverse, the LED will not turn on.

Summary

In this chapter, we learned what sensors are and their applications in electronic projects. This is important because we will continue applying sensors in other chapters from this book. We also defined photoresistors, and how they are classified. We also learned how to connect a photoresistor sensor module to the Blue Pill and Curiosity Nano boards' input ports, and how to analyze and use analog data from a photoresistor.

In this chapter, you learned important pieces of information. We have defined what a sensor and a photosensor is. You can now read data from them using microcontroller boards. The chapter also described how to connect a photosensor module to a microcontroller board.

Chapter 5, Humidity and Temperature Measurement, will explain what a humidity and temperature sensor is and how can we acquire and use its analog data with both the Blue Pill and Curiosity Nano microcontroller boards. This can have a number of applications, such as measuring the temperature and humidity of a greenhouse.

Further reading

- Horowitz, P., Hill, W. (2015), *The art of electronics* [3rd ed.], Cambridge University Press: New York, NY.

- Microchip (2019), *PIC16F15376 Curiosity Nano Hardware User Guide*, Microchip Technology, Inc. Available from: `http://ww1.microchip.com/downloads/en/DeviceDoc/50002900B.pdf`

- Mims, F.M. (2000), *Getting started in electronics*, Lincolnwood, IL: Master Publishing, Inc.

5
Humidity and Temperature Measurement

This chapter describes how to practically measure humidity and temperature in an environment, as well as how to connect specialized sensors to a microcontroller board. You will find out how to use the commonly used sensors DHT11 and LM35. In this chapter, you will gain valuable information regarding gaining data acquisition from a temperature and humidity sensor and a sensor module and how to display it to a user.

In this chapter, we are going to cover the following main topics:

- Introducing the DHT11 humidity and temperature sensor module
- Connecting the DHT11 and LM35 sensors to the microcontroller boards
- Coding to get data from the sensor module
- Showing the humidity and temperature data results on the serial port monitor

By the end of this chapter, you will have learned how to properly connect a DHT11 humidity and temperature sensor and an LM35 temperature sensor to the Curiosity Nano and Blue Pill microcontroller boards. You will have also learned how to analyze and display the data obtained from these sensors.

Technical requirements

The software tools that you will be using in this chapter are the *MPLAB X* and *Arduino* IDEs, for editing and uploading your programs to the Curiosity Nano and the Blue Pill microcontroller boards, respectively.

The code that will be used in this chapter can be found in this book's GitHub repository: `https://github.com/PacktPublishing/DIY-Microcontroller-Projects-for-Hobbyists/tree/master/Chapter05`

The Code in Action video for this chapter can be found here: `https://bit.ly/2UiRHVP`

In this chapter, we will be using the following pieces of hardware:

- A solderless breadboard.
- The Blue Pill and Curiosity Nano microcontroller boards.
- A Micro USB cable for connecting your microcontroller boards to a computer.
- The ST-LINK/V2 electronic interface needed for uploading the compiled code to the Blue Pill. Bear in mind that the ST-LINK/V2 requires four female-to-female DuPont wires.
- Green and yellow LEDs.
- Two 220-ohm resistors rated at one-quarter watt. These resistors are for the LEDs.
- One 4.7 kilo-ohm resistor rated as one-quarter watt. It is for the DHT11 sensor.
- One 2.2 kilo-ohm resistor rated as one-quarter watt. It is for the **Liquid Crystal Display (LCD)**.
- One DHT11 humidity and temperature sensor module.
- One LM35 temperature sensor.
- Three male-to-female DuPont wires for connecting the sensor module to the solderless breadboard.
- A dozen male-to-male DuPont wires.
- A 1602 16x2 LCD display.

The next section provides a brief introduction to the DHT11 sensor module, its electrical characteristics, and how this module is used in electronic projects.

Introducing the DHT11 humidity and temperature sensor module

In this section, we'll review the DHT11 sensor module. This section also describes what the sensor pins are, and how to connect them to a microcontroller board. The DHT11 is an easy-to-use, practical, and low-cost sensor module that measures temperature within a range of 0 to 50 degrees Celsius, with an error rate of +-2%. It also measures environmental **relative humidity** (**RH**) within a range of 20% to 90%, with an accuracy of +-5%. These values can change a bit, depending on the sensor module's manufacturers. The next section describes what RH is, an environmental value that is read by the DHT11 sensor.

What is relative humidity?

Relative humidity is based on a combination of water vapor and the temperature of the environment. It is a ratio of the amount of water vapor present in the air at a certain temperature, expressed as a percentage. Generally, the amount of water vapor has a higher RH in cool air than hot or warm air. A related parameter is the **dew point**, which is the temperature that air from a place or environment must be cooled down to for it to become saturated with water vapor.

Measuring RH is important because it is related to the degree of discomfort people feel in an environment, among other applications. RH measurements can have useful applications, such as in greenhouses, where some plants need a certain degree of RH to thrive.

The DHT11 pins and their values

The following is a Fritzing diagram of the DHT11 sensor module. Please note that the DHT11 module may have a fourth pin, depending on its manufacturer. The fourth pin is not connected to anything (sometimes, it is labeled **not connected**, or **NC**):

Figure 5.1 – The DHT11 sensor module

As we can see, the DHT11's **v** pin (it can also be labeled **VCC** or **+**, depending on the manufacturer) can be connected to a voltage supply of 3 to 5.5 V of **direct current** (**DC**). In this chapter, we will connect the DHT11 to a voltage of 3.3 volts, which can be supplied by both the Curiosity Nano and the Blue Pill boards. Pin **s** is the signal that provides the temperature and humidity data generated by the sensor. It can also be labeled as **OUT** or **DATA**, depending on the DHT11 manufacturer. Pin **g** (sometimes labeled as **G**, **GND**, or **-**) will be connected to the microcontroller board's ground. Most of the DHT11 modules' operating current is 0.3mA when measuring temperature and humidity, and 60 microamperes when in standby, making it a very low power sensor.

The following image shows two DHT11 sensor modules made by two different manufacturers:

Figure 5.2 – DHT11 sensor modules

Please note that the DHT11 modules have slight differences. For example, both DHT11 modules have a different pin order. The one on the left has ground (labeled as GND), signal (labeled as data), and voltage (labeled as VCC) pins in that order, whereas the right one has signal (data), voltage, and ground pins in that order. Also, the pins have different labels. However, both DHT11 modules work the same and can be used in the circuits shown in this chapter. The sensor itself is encased in the plastic blue box with a grid. The DHT11 modules have extra electronic components that facilitate the connections to the sensor.

In this section, you learn about the DHT11 sensor module, its pins, and the similarities and differences among DHT11 sensors made by different manufacturers. You also reviewed what relative humidity is, an important environmental value that the DHT11 can read.

The next section will show you how to connect the DHT11 module to a Blue Pill's digital port so that you can read its values.

Connecting a DHT11 sensor module to the microcontroller boards

This section deals with all the hardware components in this chapter. We will begin by connecting the DHT11 to the Blue Pill board. Connecting the DHT11 to microcontroller boards is easy because it only requires three wires.

Connecting the DHT11 to the Blue Pill board

In this section, we will connect the DHT11 to the Blue Pill, as shown here:

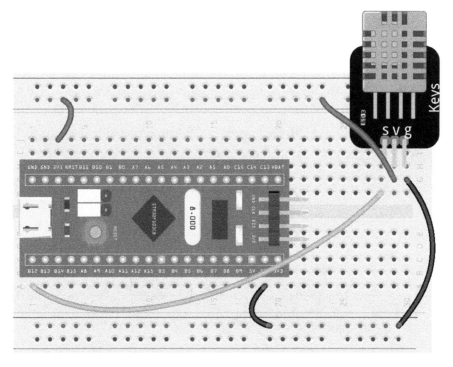

Figure 5.3 – The DHT11 connected to a Blue Pill microcontroller board

As we can see, the DHT11 connections are simple. In some DHT11 modules, their manufacturers recommend connecting a 5K-ohm pull-up resistor to the **S** (signal) pin. However, since 5K resistors are not commercially available, a resistor with a value of 4.7K-ohm is close enough to the recommended one. We did not need to connect a 4.7K-ohm to the DHT11 module that we used in our project because its electronic circuit already had a resistor like that.

> **Note**
>
> Depending on the manufacturer, many DHT11 modules already include a pull-up resistor, so the 4.7k-ohm pull up resistor is not necessary. It is worth checking it out. Just connect the DHT11 to the microcontroller board. If it gives erratic temperature measurements, or not getting measurements at all, you may need to connect the pull-up resistor to it.

Follow these steps to connect the DHT11 to the Blue Pill:

1. Connect the Blue Pill's **GND** (also labeled as **G**) pin to the solderless breadboard rail.

2. Connect the Blue Pill's **3.3** pin labeled (providing 3.3 volts) to the upper breadboard rail. Please note that in some Blue Pill boards, this pin is labeled as **3V3**.

3. Connect the DHT11's **s** pin to the Blue Pill's **B12** pin.

4. Connect the DHT11's **v** pin to the upper breadboard rail that was connected to the **3V3** pin.

5. Connect the HDT11's **g** pin to the ground pin (the lower breadboard rail).

6. Connect the USB cable to the Blue Pill and then to your computer.

The following image shows how everything should be connected:

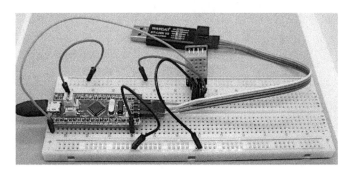

Figure 5.4 – The DHT11 connected to the Blue Pill

Here, we can see that we need just a few DuPont wires to connect the DHT11 to the Blue Pill board. You will need to connect the ST-LINK/V2 interface to your computer to upload its code, as explained in *Chapter 1, Introduction to Microcontrollers and Microcontroller Boards*. Don't forget to disconnect the USB cable first (shown on the left of the preceding image) before uploading a program to the Blue Pill. Please note that the order of the pins shown in the preceding image is **GND, DATA,** and **VCC**, which is different from the DHT11's pin order shown in *Figure 5.3*. Again, this is because some DHT11 manufacturers change the DHT11's pin order.

> **Note**
>
> All the Blue Pill's **ground** (**GND**) pins are the same; they are internally connected. This microcontroller board has more than one GND pin so that we can connect the electronic components to the board.

The temperature and humidity values will be shown on the computer's serial port monitor from the IDE. This will be explained in the *Programming the DHT11 sensor for the Blue Pill board* section.

The following image shows a Fritzing diagram with the DHT11 and a 4.7k-ohm pull-up resistor connected to the signal (**S**) pin. The DHT11 module used in this example does not have a built-in pull up resistor, so we need to connect one:

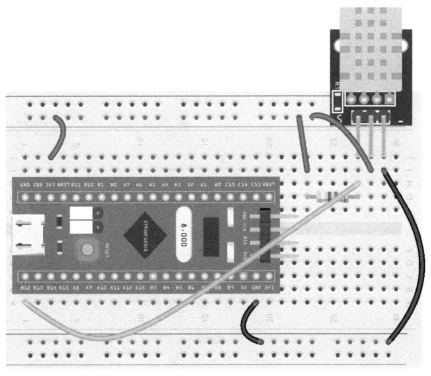

Figure 5.5 – A DHT11 with a pull-up resistor

The preceding image shows the 4.7 k-ohm resistor working as a pull-up resistor. The next section describes how to connect an LCD to show the temperature and humidity data obtained from the DHT11.

Connecting an LCD to the Blue Pill

This section shows an interesting and useful way of showing temperature and humidity data obtained from the DHT11 sensor, displaying it on a low-cost 1602 LCD. The following is a Fritzing diagram showing all the connections:

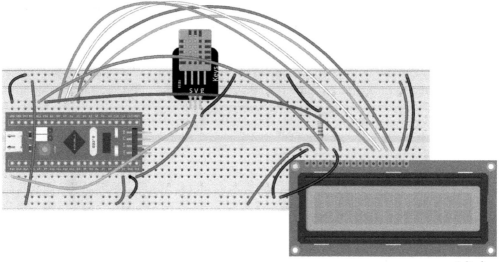

Figure 5.6 – A 1602 LCD connected to the Blue Pill board

The preceding image depicts the 1602 LCD, which can display 16 alphanumeric and special characters (for example, %, $, and so on) in two rows (16x2). Each character is formed with a 5x8 dot matrix. The LCD has 16 pins, labeled from left to right as **VSS**, **VDD**, **V0**, **RS**, **RW**, **E**, **D0**, **D1**, **D2**, **D3**, **D4**, **D5**, **D6**, **D7**, **A**, and **K**. This is a popular LCD that can also be used with Arduino microcontroller boards. The 2.2k-ohm resistor, which is connected to the **V0** LCD pin, adjusts the LCD's contrast. Pins **VSS** and **VDD** are connected to a power supply. Pins **D0** to **D7** are used to send data to the LCD to form characters, but here, we're only using 4 bits (pins **D4** to **D7**) to display the letters and numbers on the LCD. The 1602 LCD is connected to a **5-volt** (**+5V**) power supply, which is supplied by the Blue Pill's 5V pin.

Bear in mind that the upper and lower breadboard voltage rails shown in the preceding image are connected to Blue Pill's 5V pin, providing 5 volts to the LCD. The DHT11 is connected to the Blue Pill's 3.3 pin because this sensor works with 3.3 volts. Its resulting data will be sent to the Blue Pill's input data pin, **B12**, using a digital voltage level with 3.3 volts.

Follow these steps to connect the DHT11 and the LCD to the Blue Pill while following the preceding image:

1. Connect the Blue Pill's **GND** (also labeled as **G**) pins to the solderless breadboard rails.

2. Connect the Blue Pill's **5V** pin (providing 5 volts) to the breadboard rails.

3. Connect the DHT11's **s** pin to the Blue Pill's **B12** pin.

4. Connect the DHT11's **v** pin to the Blue Pill's **3V3** pin.

5. Connect the HDT11's **g** pin to the ground pin (the upper breadboard rail).

6. Connect the USB cable to the Blue Pill and then to your computer or a USB power bank.

7. Insert the LCD's 16 pins into the solderless breadboard.

8. Connect the LCD's **VSS** pin to the ground pin (the lower breadboard rail).

9. Connect the LCD's **VDD** pin to 5 volts (the lower breadboard rail).

10. Connect the 2.2K-ohm resistor to the LCD's **V0** pin and to the ground pin (the upper breadboard rail).

11. Connect the LCD's **RS** pin to the Blue Pill's **B11** pin.

12. Connect the LCD's **RW** pin to the ground pin (lower breadboard rail).

13. Connect the LCD's **E** pin to the Blue Pill's **B10** pin.

14. Connect the LCD's **D4** pin to the Blue Pill's **B0** pin.

15. Connect the LCD's **D5** pin to the Blue Pill's **A7** pin.

16. Connect the LCD's **D6** pin to the Blue Pill's **A6** pin.

17. Connect the LCD's **D7** pin to the Blue Pill's **A5** pin.

18. Connect the LCD's **A** pin to 5 volts (upper breadboard rail).

19. Connect the LCD's **K** pin to the ground pin (upper breadboard rail).

20. The LCD's **D0**, **D1**, **D2**, and **D3** pins are not connected.

The following image shows how everything is connected:

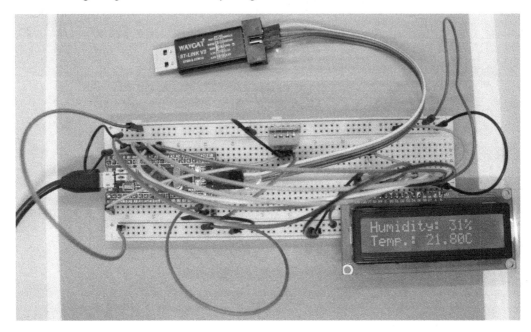

Figure 5.7 – The 1602 LCD connected to the Blue Pill microcontroller board

Here, you can see the LCD working because the Blue Pill is connected to a USB power bank (not shown in the preceding image). The LCD is displaying the local temperature in degrees Celsius, and the humidity is shown as a percentage. The code for programming the LCD and the DHT11 will be shown later in this chapter.

The next section will show you how to connect an LM35 temperature sensor to the Curiosity Nano board, demonstrating how to read temperature values from that sensor and sending them to an analog input port from the Curiosity Nano.

Connecting an LM35 temperature sensor to the Curiosity Nano board

In this section, we'll describe how to measure temperature values on a Curiosity Nano, obtained from an LM35 sensor. For demonstration purposes, the temperature will be analyzed when it falls within a thermal comfort zone of 23 to 26 degrees Celsius. The LM35 is a low-cost and easy-to-connect sensor that measures a wide temperature range, from -55 to 150 degrees Celsius. The following is a diagram of the LM35 sensor pinout:

Figure 5.8 – The LM35 sensor showing pins 1, 2, and 3

As you can see, the LM35 has three pins: pin 1 (**Vs**), pin 2 (**Vout**), and pin 3 (**GND**). Pin 1 is the voltage coming from a power supply. The Curiosity Nano's **VBUS** pin provides 5 volts, so the LM35's pin 1 can be connected to it. Pin 2 is voltage out (**Vout**), providing the measured temperature value in terms of **millivolts** (**mV**) using a linear scale factor. This scale increases+10 mV for each degree Celsius (+10 mV/C). Pin 3 is connected to the microcontroller board's ground (**GND**).

The LM35 sensor signal pin (pin 2) provides the temperature values in mV, which will be converted into digital values by the Curiosity Nano. To do this, we need to connect the LM35's pin 2 to an input analog port from the Curiosity Nano. The temperature value is calculated as *temperature=(5.0/1023)*millivolts_from_LM35*, where 5.0 is the 5 volts connected to the LM35's pin 1 and 1023 is the 10-bit ADC from the Curiosity Nano, which means that its ADC can detect (2^10) discrete analog levels. The formula converts the LM35's output, in mV, into an equivalent value in degrees Celsius. The following image shows the LM35 sensor:

Figure 5.9 – The LM35 temperature sensor

The preceding image shows the LM35 temperature sensor with its three pins. It is the same size as a regular transistor since both have a TO-92 semiconductor package type. However, the LM35 is not a transistor – it's a temperature sensor.

We must connect two LEDs to its output ports to show the temperature and humidity ranges, as shown here:

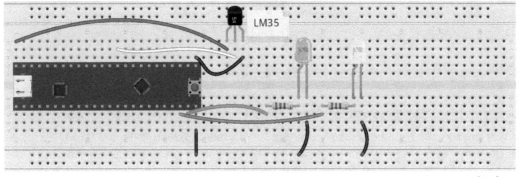

Figure 5.10 – Connecting the LM35 sensor and the LEDs to the Curiosity Nano board

As you can see, we must connect a green LED to show that the environment temperature being read by the LM35 sensor is within the thermal comfort zone for an office with a range of 23 to 26 degrees Celsius; and a yellow LED will be turned on if the temperature is outside that range. The temperature values from the thermal comfort zone used in this chapter are determined by the Canadian Standards Association, which are similar to the values defined by other agencies from other countries. The thermal comfort zone values can be found here: `https://www.ccohs.ca/oshanswers/phys_agents/thermal_comfort.html`.

Follow these steps to connect the LM35 and the LEDs to the Curiosity Nano board:

1. Connect the Curiosity Nano's **GND** pin to the lower rail of the solderless breadboard rail.

2. Connect the LM35's **Vout** (signal) pin (2) to the Curiosity Nano's **RA0** pin. It will be used as an analog input port.

3. Connect the LM35's **Vs** pin (1) to the Curiosity Nano's **VBUS** pin. This pin provides 5 volts.

4. Connect the LM35's **g** pin (3) to the Curiosity Nano's **GND** pin from its upper pin row.

5. Now, connect the two 220-ohm resistors to the Curiosity Nano's **RD2** and **RD3** ports and to the anodes of the two LEDs, respectively.

6. Connect the LEDs' cathodes to the ground pin.

7. Finally, connect the USB cable to the Curiosity Nano and then to your computer or a USB power bank.

The following image shows how everything should be connected:

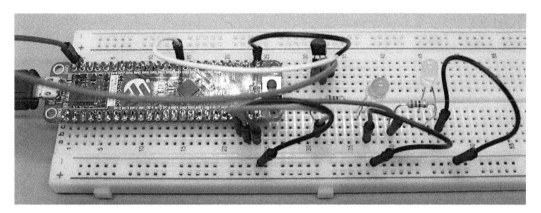

Figure 5.11 – The LM35 sensor connected to the Curiosity Nano board

The preceding image shows the LM35 connected to the Curiosity Nano's RA0 at the upper pins row.

> **Note**
> All the Curiosity Nano's GND pins are the same; they are internally connected. This microcontroller board has more than one GND pin so that we can connect electronic components to the board.

In this section, you learned about the LM35 temperature sensor, its pins, and how to connect it to a microcontroller board. You also learned how to show the temperature information with LEDs. The next section will show you how to write programs for reading humidity and temperature values from the DHT11 sensor.

Coding to get data from the sensor module

This section describes how to code a program for both the Blue Pill and the Curiosity Nano boards so that they can read values from the DHT11. You will also learn how to program the LM35 temperature sensor with the Blue Pill. Let's start by programming the DHT11 sensor for the Blue Pill board.

Programming the DHT11 sensor for the Blue Pill board

In this section, you will review the code that gets data from the DHT11 sensor using a special library. The code will also show you how to display the temperature and humidity data on the serial port, and thus on the Arduino IDE's serial monitor. The following code reads both the temperature and humidity from a DHT11 sensor module, which is connected to digital input port B12 of the Blue Pill:

```
#include <DHT.h>
#define DHT11_data_pin PB12
DHT dht(DHT11_data_pin, DHT11);
void setup() {
    Serial.begin(9600);
    while (!Serial);
    Serial.println("Opening serial comm.");

    dht.begin();
}

void loop() {
    float humidity = dht.readHumidity();
    float temperature=dht.readTemperature();
    Serial.println("Humidity: "+String(humidity));
    Serial.println("Temperature: "+String(temperature));
    delay(1000);
}
```

In the preceding code, you can see that the first line includes a library called DHT.h. This is a very practical library for reading values from the DHT11. This code can be found in this book's GitHub repository, which contains useful comments explaining its main parts. To install the DHT.h library on the Arduino IDE, follow these steps:

1. Click on **Tools | Manage Libraries** from the IDE's main menu.

2. Type DHT11 in the Library Manager's search field.

3. Several libraries will be listed. Select and install the highest version of the DHT sensor library, made by Adafruit.

4. Wait until the library has been installed. Then, close the Library Manager. Now, the DHT11 library should be ready to use in your code.

Please note that the dht.readTemperature() function returns temperature values in degrees Celsius, while the dht.readHumidity() function returns values as a percentage of RH.

The following screenshot shows the Arduino IDE's **Library Manager** showing the library called **DHT sensor library by Adafruit**:

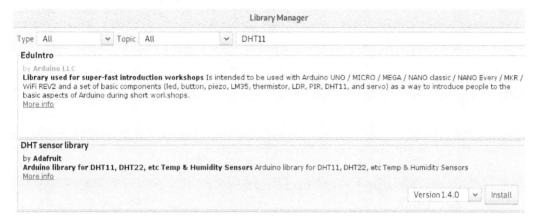

Figure 5.12 – The Arduino IDE's Library Manager

Here, you can see that by typing DHT11 in the search field, you can find the DHT sensor library by Adafruit. Click on the **Install** button.

> **Tip**
> You can also connect a DHT11 sensor to an Arduino Uno microcontroller board. Just connect the DHT11's **s** (signal) pin to any digital port from 2 to 13 from the Arduino Uno and change the pin number in the constant DHT11_ data_pin shown in the preceding code. In addition, connect the DHT11's GND and VCC pins to the Arduino Uno's GND and 3V3 pins, respectively.

The next section will show you how to display the humidity and temperature data on a 1602 LCD.

Coding the sensor module and the 1602 LCD

The following code describes how to get the temperature and humidity data from the DHT11 sensor and how to display that data on the 1602 LCD:

```
#include <DHT.h>
#include <math.h>
#include <LiquidCrystal.h>
```

```
const int rs = PB11, en = PB10, d4 = PB0, d5 = PA7,
    d6 = PA6, d7 = PA5;
LiquidCrystal lcd(rs, en, d4, d5, d6, d7);
#define DHT11_data_pin PB12
DHT dht(DHT11_data_pin, DHT11);
void setup() {
    dht.begin();
    lcd.begin(16, 2);
}

void loop() {
    float humidity = dht.readHumidity();
    float temperature=dht.readTemperature();
    lcd.setCursor(0, 0);
    lcd.print("Humidity: "+String(round(humidity))+"%    ");
    lcd.setCursor(0,1);
    lcd.print("Temp.: "+String(temperature)+"C    ");
    delay(1000);
}
```

Bear in mind that the code that's been uploaded to this book's GitHub repository contains many comments explaining the main parts of the preceding code. `LiquidCrystal.h` is the library that's used for controlling the 1602 LCD. The `math.h` library contains the `round()` function, which is used for rounding the humidity value. Both libraries belong to the Arduino IDE's original installation files, so we don't need to download or install them separately. `DHT.h` is a library for reading DHT11 values; its installation was explained previously in this chapter.

The next section describes how to code the LM35 temperature sensor to the Curiosity Nano microcontroller board.

Programming the LM35 sensor on the Curiosity Nano board

The following code reads the temperature from an LM35 sensor connected to analog input port RA0 of the Curiosity Nano:

```
#include "mcc_generated_files/mcc.h"
static uint16_t LM35read=0;
```

```
float temp=0.0;
void main(void)
{
     // initialize the device
     SYSTEM_Initialize();
     ADC_Initialize();
     while (1)
     {
          IO_RD2_SetLow();
          IO_RD3_SetLow();
          LM35read=ADC_GetConversion(channel_ANA0);
          temp=(5.0/1023)*LM35read;
          if (temp>=23.0 && temp<=26.0)
          {
               IO_RD3_SetHigh();
          } else {
            IO_RD2_SetHigh();
          }
          __delay_ms(500);
     }
}
```

As you can see, the Curiosity Nano performs an analog to digital conversion for the value that is read from the LM35 sensor using the `ADC_GetConversion()` function. We also used the formula for calculating the degrees Celsius according to the mV read from the LM35; that is, `temp=(5.0/1023)*LM35read;`.

With that, you have learned how to get data from the DHT11 sensor module and the LM35 sensor by coding the Blue Pill and Curiosoty Nano. The next section will show you how to display the DHT11 data on the Arduino IDE's serial monitor.

Showing the humidity and temperature data results on the serial port monitor

This section describes how the DHT11 data is displayed with the Blue Pill on the serial port monitor and the Curiosity Nano microcontroller boards, and also discusses what to do next with the data that's obtained from the DHT11 sensor module.

Open and run the code for the Blue Pill on the Arduino IDE. You can read the data that was obtained from the DHT11 by clicking on **Tools | Serial Monitor** from the Arduino IDE's main menu. This can be seen in the following screenshot:

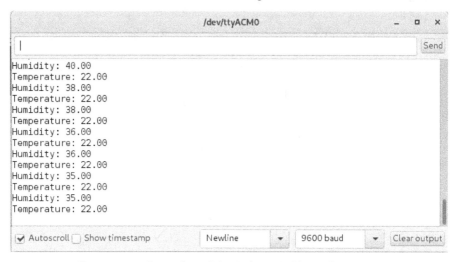

Figure 5.13 – Screenshot of the Arduino IDE's serial monitor

Here, you can see that the humidity and temperature values that were obtained from the DHT11 are shown with decimal points, thus showing a more accurate temperature. Remember that the variables storing these values were declared with the `float` type.

> **Tip**
> If you can't open the IDE's serial monitor and if the IDE shows a message telling you that the USB port cannot be found, it may be that the USB cable that you connected to the Blue Pill and to your computer is faulty. In addition, some USB cables are not capable of transmitting data because those cables are only used for charging devices, which you should not use with the Blue Pill.

Please note that the temperature shown in the preceding screenshot is in degrees Celsius and that the RH is displayed as a percentage.

Plotting the data

The Arduino IDE's serial port monitor has an interesting option for plotting the values that are read from the DHT11 and sent to the serial port graphically. Those values are plotted in real time.

Click on **Tools | Serial Plotter** from the Arduino IDE's main menu and see how the data is displayed graphically, as shown here:

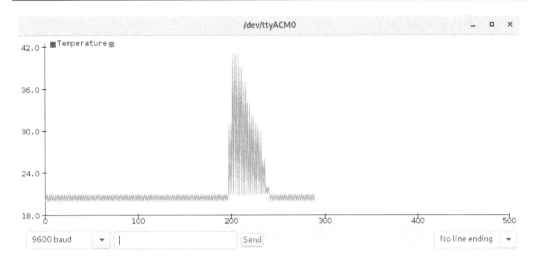

Figure 5.14 – Screenshot of the HDT11 serial plotter

Here, you can see how the DHT11 data is plotted graphically. The *y* axis represents the amount of humidity, while the *x* axis represents the time in seconds. The graph shows a peak because the environmental humidity changed at that time. To test how the DHT11 sensor's humidity measures change, just blow on the sensor.

In this section, you learned how to open the Arduino IDE's serial monitor and how the DHT11 sensor module data is displayed on it, including how to plot that data graphically.

Summary

In this chapter, we learned the basics of the DHT11 humidity and temperature sensor and the LM35 temperature sensor, including their operating ranges and how these sensors send data to a microcontroller board. These are practical and low-cost sensors that the Blue Pill and the Curiosity Nano can easily handle. This chapter showed two ways of showing humidity and temperature results. The Blue Pill showed the humidity and temperature values directly on the computer monitor and on an LCD, while the Curiosity Nano showed temperature and humidity ranges with LEDs. This chapter was beneficial because you learned how to get environmental data from sensors using microcontroller boards and how to display it effectively, using the microcontroller boards' input and output ports. This chapter also highlighted the use of an LCD to show the temperature and humidity data.

Chapter 6, Morse Code SOS Visual Alarm with a Bright LED, will review how to create a practical visual alarm.

Further reading

- Horowitz, P., Hill, W. (2015). *The art of electronics*. [3rd ed.] Cambridge University Press: New York, NY.

- LM35 (2017). LM35 Precision Centigrade Temperature Sensor datasheet. Texas Instruments. Available from: `https://www.ti.com/lit/ds/symlink/lm35.pdf`

- Microchip (2019). *PIC16F15376 Curiosity Nano hardware user guide*. Microchip Technology, Inc. Available from: `http://ww1.microchip.com/downloads/en/DeviceDoc/50002900B.pdf`

- Mouser (2020). *DHT11 humidity and temperature sensor datasheet*. Mouser, Inc. Available from: `https://www.mouser.com/datasheet/2/758/DHT11-Technical-Data-Sheet-Translated-Version-1143054.pdf`

6
Morse Code SOS Visual Alarm with a Bright LED

This chapter describes how to build a very noticeable visual alarm with a **super-bright LED** connected to a microcontroller board. The LED will show an SOS message (this is used when someone or a group of people is in danger or in distress). In this chapter, you will learn how to control a super-bright LED with a microcontroller board. The reason we are using a super-bright LED in this chapter is to increase the visibility of the SOS message with it. This chapter will be beneficial for future electronic projects because you will learn how to control an LED with a transistor working as a switch. The application made in this chapter can be used by people in distress while they are hiking, at sea, and in similar scenarios.

In this chapter, we are going to cover the following main topics:

- Understanding Morse code and the SOS message
- Introducing super-bright LEDs and calculating their necessary resistors

- Connecting the resistor and the super-bright LED to the microcontroller board
- Coding the SOS Morse code signal
- Testing the visual alarm

By the end of this chapter, you will be able to properly connect a super-bright LED to the Curiosity Nano and Blue Pill microcontroller boards and to generate a Morse code message with a microcontroller board.

Technical requirements

The software tools that you will be using in this chapter are the MPLAB-X and Arduino IDEs for editing and uploading your programs to the Curiosity Nano and the Blue Pill microcontroller boards, respectively.

The code used in this chapter can be found at the book's GitHub repository:

`https://github.com/PacktPublishing/DIY-Microcontroller-Projects-for-Hobbyists/tree/master/Chapter06`

The Code in Action video for this chapter can be found here: `https://bit.ly/3iXDlEP`

In this chapter, we will use the following pieces of hardware:

- A solderless breadboard.
- The Blue Pill and Curiosity Nano microcontroller boards.
- A micro USB cable for connecting your microcontroller boards to a computer.
- The ST-Link/V2 electronic interface needed to upload the compiled code to the Blue Pill. Bear in mind that the ST-Link/V2 requires four female-to-female DuPont wires.
- One 5 mm trough-hole 625 nm orange-red, super-bright LED, manufacturer number BL-BJU334V-1, made by American Bright Optoelectronics Corporation, or something similar.
- Two 1 k ohm resistors rated at one-quarter watt. These resistors are for the LED and the transistor.
- One 220-ohm resistor rated at one-quarter watt. This is for the LED.
- One **2N2222** transistor, TO-92 package.
- A dozen male-to-male DuPont wires.

The following section describes what Morse code is and why we are using it in this project.

Understanding Morse code and the SOS message

Morse code is a telecommunication technique used for encoding, receiving, and sending **alphanumeric** and **special characters** by applying signal sequences with different duration. Morse code is named after Samuel Morse, a telegraph inventor. This code is important because it was commonly used for radio and wired communication over long distances, in particular, for sending and receiving telegrams. Nowadays, Morse code is still used in amateur (*ham*) radio communications because it can be reliably decoded by people when electromagnetic atmospheric conditions are unfavorable. More importantly, Morse code can be used in an emergency by sending SOS messages in the form of light, audio, or electromagnetic signals. Morse code is still sometimes used in aviation as a radio navigation aid.

Each character in Morse code is made with **dashes** (represented by the - symbol) and **dots** (represented by the . symbol). A dot is one signal unit, and a dash is three signal units. Each alphabet letter, numerals, and special characters are encoded using a combination of dots and/or dashes. A space between letters is one signal unit (a dot), and a space between words is seven signal units. This code is generally transmitted on an information-carrying medium, for example, visible light and electromagnetic radio waves. The following is a list with the letters and numbers encoded using international Morse code:

- A (. -)
- B (- . . .)
- C (- . - .)
- D (- . .)
- E (.)
- F (. . - .)
- G (- - .)
- H (. . . .)
- I (. .)
- J (. - - -)
- K (- . -)
- L (. - . .)
- M (- -)

- N (-.)
- O (- - -)
- P (.- -.)
- Q (- -.-)
- R (.-.)
- S (. . .)
- T (-)
- U (..-)
- V (...-)
- W (.- -)
- X (-..-)
- Y (-.--)
- Z (- -..)
- 1 (.- - - -)
- 2 (..- - -)
- 3 (...- -)
- 4 (. . . . -)
- 5 (. . . . .)
- 6 (-. . . .)
- 7 (- - . . .)
- 8 (- - - . .)
- 9 (- - - -.)
- 0 (- - - - -)

Special characters, such as $ and # and other characters from languages other than English, have also been encoded with Morse code, but showing them is beyond the scope of this chapter. For example, this is how we encode the word PARIS in Morse code:

.- -. .- .-. .. ...

This is another example. The word HELP can be codified as .... . .-.. .--.

Another longer example is the word ALIVE: .-   .-..   ..   ...-   .

A commonly used distress message encoded with Morse code is made up of the letters SOS. It is formed by three dots, three dashes, and three dots (.   .   .   - - -   .   .   .). This message has been internationally applied and recognized by treaties, originally used for maritime emergency situations. Its origin is uncertain, but popular usage associates SOS with phrases such as *Save Our Ship* or *Save Our Souls*. Nonetheless, SOS is easy to remember and used in an emergency and is shorter than coding other words, such as *HELP*, in Morse code.

In this chapter, we will use the SOS message to make a visual alarm, showing that message making the dots and dashes by turning on and off a super-bright LED with the Blue Pill and the Curiosity Nano boards.

The next section describes a brief introduction to super-bright LEDs, and what type of resistor can be connected to them for our Morse code purposes.

Introducing super-bright LEDs and calculating their necessary resistors

A **super-bright LED** is a **light-emitting diode** (**LED**) that glows with high intensity, higher than regular LEDs. LED brightness (light intensity) is calculated in **millicandelas** (**mcd**). Bear in mind that 1,000 mcd equals 1 candela. Candelas typically measure how much light is generated at the light source, in this case, an LED, but candelas can be used to measure other light sources, such as light bulbs. The super-bright LED that we use in this chapter is rated as 6,000 mcd, emitting a nice and powerful orange glow, which is quite bright when connected to a proper current-limiting resistor. In comparison, typical LEDs are rated at a range of about 50 to 200 mcd.

Super-bright LEDs have a special design to increase light diffusion by using a transparent glass coating and reflective material. However, some super-bright LEDs have a reduced **viewing angle** (the observation angle with the LED light looks more intense) of about 35 degrees, such as the one we are using in this chapter, whereas the viewing angle of other regular and super-bright LEDs is 120%. This viewing angle depends on their cost, efficiency, and applications.

As with regular LEDs, super-bright LEDs require a certain voltage to power them, typically between 2 and 3 volts. That's why we need to connect a current-limiting resistor to an LED to reduce its voltage. We can use the formula R=(Vs-Vf)/If for calculating a current-limiting resistor for an LED in the following cases:

- **Vs** = supplied voltage. The Blue Pill and Curiosity Nano output ports provide 3.3 V.

- **Vf** = forward voltage, which is the voltage that drops through a resistor.

- **If** is the forward **amperage (amps)**.

- **R** is the resistor value that we want to calculate.

A commonly used resistor value for connecting to LEDs is 1 k ohm. The drop voltage with that resistor is 1.8 V when applying a supply voltage of 3.3 volts and a current drawing 1.5 milliamps (or 0.0015 amps). Let's apply the preceding formula to confirm this resistor value:

$$R = (Vs - Vf)/Is = (3.3 - 1.8)/0.0015 = 1000 ohms, \lor 1 kohm.$$

The resistor connected to a super-bright LED will determine the number of amps it will draw. Typical resistor values that are used for connecting to LEDs are 220, 330, and 1 k ohms when using either 3.3 V or 5 V as the supply voltage that many microcontroller boards supply. The resistor value is not critical in the majority of LED applications (unless you are connecting an LED to a Blue Pill microcontroller board, as you will see in the next section), so we can use other resistors with similar values. First of all, you can determine whether the super-bright LED works. Go to the *Testing the visual alarm* section at the end of this chapter.

> **Important note**
>
> Do not stare at a super-bright LED directly on top of it when it is turned on (glowing), as this may hurt your eyes. You can momentarily look at it from the side.

Figure 6.1 shows the super-bright LED used in this chapter:

Figure 6.1 – The BL-BJU334V-1 super-bright LED

Figure 6.1 depicts a BL-BJU334V-1 5 mm through-hole LED that emits bright orange light in the 625 nm wavelength. Its left pin is the anode (positive lead), and the right pin is the cathode (negative lead) which is the short one. There are other types of super-bright LEDs with higher brightness and sizes. We decided to use this one in particular for this chapter because is low cost and suitable for inserting it in a solderless breadboard and connecting it to a microcontroller board. The next section deals with the connection of a super-bright LED to the Blue Pill and Curiosity Nano boards.

Connecting the resistor and the super-bright LED to the microcontroller board

This section shows how to use a super-bright LED connected to a microcontroller board to display a Morse code message. We begin by explaining how to connect a super-bright LED to one of the input ports of the Blue Pill and how to use a transistor as a switch to control the super-bright LED. Then, we describe how to connect the super-bright LED to the Curiosity Nano board.

Figure 6.2 shows a Fritzing diagram containing a super-bright LED:

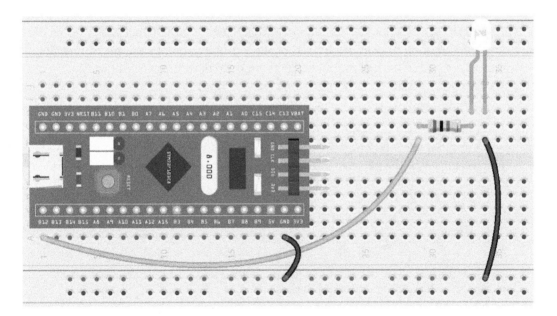

Figure 6.2 – A super-bright LED connected to a Blue Pill's I/O port

As you can see from *Figure 6.2*, the super-bright LED's anode is connected to a 1 k ohm current-limiting resistor. The resistor is connected to output port B12, providing 3.3 V to it every time a dot or dash from a Morse code character is sent to it. The following are the steps for connecting all the components shown in *Figure 6.2*:

1. Connect the 1 k ohm resistor to the Blue Pill's B12 port.

2. Connect the resistor to the super-bright LED's anode pin (its longest leg, which is the left one shown in *Figure 6.1*).

3. Connect the super-bright LED's cathode pin (its shortest leg) to the lower solderless breadboard's row ground.

4. Connect the Blue Pill's GND pin to the solderless breadboard's lower rail.

When doing those steps, be sure to connect the right LED polarity.

> **Important note**
>
> Do *not* connect a super-bright LED directly to the power supply, as you will damage it. You should connect a current-limiting resistor to its anode (the long LED leg).

Each Blue Pill I/O port can handle up to 6 mA (milliamps). Be careful, because in some cases and configurations, a super bright LED could consume more than 6 mA of current. That's why we connected a 1 k ohm resistor to its anode to limit the current (and its voltage) arriving at the LED. If anything connected to a Blue Pill port is drawing more than 6 mA, you will damage the Blue Pill board.

Before we connect everything to a Blue Pill microcontroller board port, we need to know how many amps in the I/O **B12** port are consumed by the LED to see whether it is below the maximum amps that the port can handle. The amps (current) consumed by the LED from *Figure 6.2* is calculated using the following formula:

$$I = (Vs - Vf)/R$$

The following are the descriptions for the symbols:

- Vs = 3.3 V, which is the supplied voltage given by the output port generated when making the Morse code's dots and dashes.

- Vf = 1.8 V, which is the forward voltage that drops across the 1 k resistor (we measured it with a multimeter).

- R is the resistor value (1 k ohm).

Thus, $I = (3.3 - 1.8)/1000 = 0.0015 amps$, or 1.5 milliamps, well below the maximum 6 mA that each Blue Pill port can handle, so we can safely use a 1 k ohm resistor for our super bright LED if we connect it to a port that produces 3.3 V.

It is important to note that the resistor value and the voltage arriving at the resistor from the port will determine the LED brightness. We could use a resistor with a lower value for making the LED glow brighter. For example, we could use a 220-ohm resistor. Its dropping voltage is 1.9 V (we measured it with a multimeter). Let's calculate its amperage (amps) using Vs = 3.3 V: I=(Vs-Vf)/R=(3.3-1.9)/220=6 mA, which is the limit amps that a Blue Pill port can handle. If we use a 220-ohm (or a lower value) resistor for connecting it to a super-bright LED, we should use a transistor working as a switch for protecting the Blue Pill port by not drawing current directly from it. *Figure 6.3* shows a Fritzing diagram with a 2N2222 transistor. It will close its *switch*, that is, connect its collector and emitter internally, when it receives a voltage to its base (transistor pin number 2), and thus connect the LED's cathode to the ground:

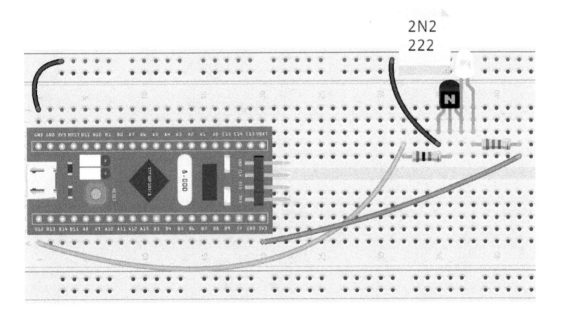

Figure 6.3 – The super-bright LED switched on/off by the 2N2222 transistor

As you can see from *Figure 6.3*, the LED is connected to a 220-ohm resistor that is connected to the Blue Pill's 3.3 V pin. This resistor draws the current from the Blue Pill's 3.3 pin, which handles up to 300 mA, and not from the B12 port handling up to 6 mA. Thus, the transistor controls the power flow to another part of the circuit switching the LED's ground. Every time the transistor receives a certain voltage at its base, it saturates and creates the binary on/off switch effect between its collector and emitter pins. We need to connect a resistor to its base to reduce its voltage and thus properly saturate it. A typical value for saturating the transistor used in this chapter is 1 k ohm when applying 3.3 V to it. The low-cost and popular transistor used in *Figure 6.3* has the part number 2N2222 (in a TO-92 package). This transistor handles up to 600 mA, enough to drive our super-bright LED. The following are the steps for connecting all the components shown in *Figure 6.3*:

1. Connect the Blue Pill's GND pin to the breadboard's upper row.
2. Connect the 1 k ohm resistor to the Blue Pill's B12 port.
3. Connect the 1 k ohm resistor to the 2N2222 transistor's base (pin number **2**).
4. Connect the 2N2222 transistor's emitter (pin number **1**) to the ground (the breadboard's upper row).
5. Connect the LED's cathode pin (its shortest leg) to the 2N2222's collector (pin number **3**).
6. Connect the 220-ohm resistor to the LED's anode pin (its longest leg).
7. Connect the 220-ohm resistor to Blue Pill's 3V3 (3.3) pin.

Figure 6.4 shows the 2N2222 pinout:

Figure 6.4 – The 2N2222 transistor pin numbers

Figure 6.4 shows a 2N2222 transistor in its TO-92 package, showing its three pins. Pin 1 is its emitter, pin 2 is its base, and pin 3 is its collector. The N shown on the transistor means that the 2N2222 is an NPN-type transistor, an internal configuration composed of negative-positive-negative layers. That big N letter is not actually shown on commercial transistors; it appears only on the Fritzing diagram shown in *Figure 6.4*. Instead, commercial transistors show their part number on them.

Figure 6.5 shows the electronic symbol for the transistor:

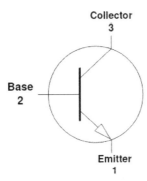

Figure 6.5 – The transistor's electronic diagram (NPN type)

As you can see from *Figure 6.5*, the symbol shows the 2N2222's pin numbers. This pin order will change in other types of transistors.

Figure 6.6 shows how everything is connected:

Figure 6.6 – Connecting the super-bright LED to a 2N2222 transistor

Figure 6.6 shows the 2N2222 transistor connected to the BluePill's ground and 3.3 (3.3 V) pins. The transistor is connected at the left of the super-bright LED. The next section deals with connecting the super-bright LED to the Curiosity Nano board using a resistor and a transistor.

Connecting the super-bright LED to the Curiosity Nano

This section explains how to control the super-bright LED from the Curiosity Nano board to display the SOS Morse code message.

Figure 6.7 is a Fritzing diagram showing how to connect a super-bright LED to the Curiosity Nano:

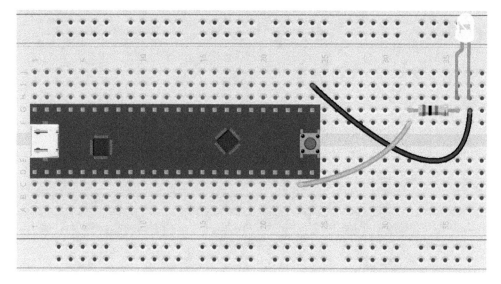

Figure 6.7 – A super-bright LED connected to a Curiosity Nano's I/O port

As you can see in *Figure 6.7*, we use a 1 k ohm resistor to connect the super-bright LED, which is a similar circuit for the Blue Pill microcontroller board explained in a previous section. Here are the steps for connecting the components shown in *Figure 6.7*:

1. Connect the Curiosity Nano's RD3 pin to a 1 k ohm resistor.
2. Connect the resistor to the super-bright LED's anode pin (its longest leg).
3. Connect the super-bright LED's cathode pin (its shortest leg) to the Curiosity Nano's GND pin.

Be aware that in theory, each of the Curiosity Nano I/O ports can handle up to 12.8 milliamps (500 milliamps/39 ports), and the Curiosity Nano's voltage regulator (VBUS) can handle up to 500 milliamps. This may change a bit with the ambient temperature, according to its manufacturer. We could make the LED glow brighter by connecting a lower-value resistor, or we could connect more than one LED to a port, but this will draw more current, potentially damaging the microcontroller board. Since each port cannot support a lot of current, we need to connect a transistor to switch the LED's ground. *Figure 6.8* shows how all the components are connected:

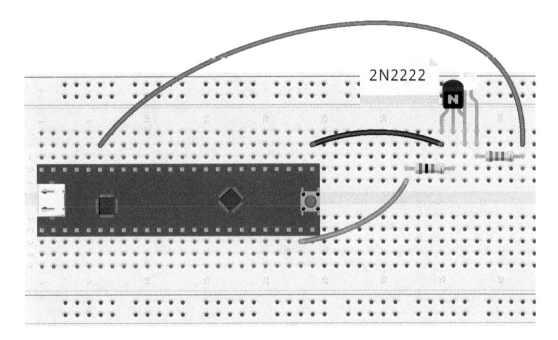

Figure 6.8 – The super-bright LED switched on/off by the transistor, connected
to the Curiosity Nano board

As you can see in *Figure 6.8*, the resistor connected to the LED's anode pin is connected to the Curiosity Nano's VTG port that provides 3.3 V. These are the steps for connecting all the components:

1. Insert the Curiosity Nano, the 2N2222 transistor, and the super-bright LED to the solderless breadboard.
2. Connect the transistor's pin 1 to the Curiosity Nano's ground (GND) pin.
3. Connect the 1 k ohm resistor to Curiosity Nano's RD3 port.
4. Connect the 1 k ohm resistor to the transistor's pin 2.
5. Connect the super-bright LED's cathode (its shortest pin) to the transistor's pin 3.
6. Connect the 220-ohm resistor to the super-bright LED's anode (its longest pin).
7. Connect the 220-ohm resistor to Curiosity Nano's VTG pin.

Figure 6.9 shows how everything is connected:

Figure 6.9 – The transistor and the LED connected to the Curiosity Nano

As you can see from *Figure 6.9*, the 2N2222 transistor is switching the super-bright LED's ground.

The next section describes how the SOS signal can be programmed on the Blue Pill and the Curiosity Nano boards.

Coding the SOS Morse code signal

This section describes the code necessary for turning the LED for showing the SOS Morse signal, which runs on the Blue Pill board, on and off. The following code shows the main functions used for defining the SOS Morse code message and for sending it to the board's output port. The next code segment defines the necessary dot, dash, and space values, as well as the port label:

```
int led=PB12;
int dot_duration=150;
int dash_duration=dot_duration*3;
int shortspace_duration=dot_duration;
int space_duration=dot_duration*7;
```

The next function sets up the output port (B12) for turning the LED on and off:

```
void setup() {
        pinMode (led,OUTPUT);
}
```

These functions define the letter S and O to be used in the SOS message:

```
void S() {
        dot();
        dot();
        dot();
        shortspace();
}

void O() {
        dash();
        dash();
        dash();
        shortspace();
}
```

The following functions define the time spaces in between the letters and the space in between each SOS message sent to the output port:

```
void shortspace() {
        delay(shortspace_duration);
}
void space() {
        delay (space_duration);
}
void dot() {
        digitalWrite(led,HIGH);
        delay (dot_duration);
        digitalWrite(led,LOW);
        delay(dot_duration);
}
```

```
void dash() {
        digitalWrite(led,HIGH);
        delay(dash_duration);
        digitalWrite(led,LOW);
        delay(dash_duration);
}
```

The following is the main function, which defines the SOS message and its leading time space:

```
void loop() {
        S(); O(); S();
        space();
}
```

The preceding code shows that the Blue Pill's B12 port is used as an output for turning the super-bright LED on and off. In this example code, each dot has a duration of 150 milliseconds (stored in the dot_duration variable), and each dash has a duration of three times that of each dot. There is a short space of time in between each letter, made by the shortspace() function. Also, there is a time space in between each SOS word, made by the space() function. The letters S and O from the SOS Morse message are encoded by the functions S() and O(), respectively.

Bear in mind that the code uploaded to the GitHub repository has many comments explaining most of the instructions from the preceding code.

> **Note**
> You can run the preceding code on an Arduino microcontroller board. Just change the output port number PB12 to any Arduino digital port. For example, the Arduino Uno microcontroller board has digital ports 2 to 13.

The following section shows how to code the SOS message example on the Curiosity Nano board.

The SOS message code for the Curiosity Nano

The SOS message code for the Curiosity Nano microcontroller board is similar to the one that runs on the Blue Pill board. The next code segment defines the necessary dot, dash, and space values:

```
#include "mcc_generated_files/mcc.h"
const int dot_duration=150;
const int dash_duration=dot_duration*3;
const int shortspace_duration=150;
const int space_duration=dot_duration*7;
```

The following functions define the time spaces in between the letters and the space in between each SOS message sent to the output port:

```
void shortspace() {
        __delay_ms(shortspace_duration);
}
void space() {
        __delay_ms(space_duration);
}
void dot() {
        IO_RD3_SetHigh();
        __delay_ms(dot_duration);
        IO_RD3_SetLow();
        __delay_ms(dot_duration);
}
void dash() {
        IO_RD3_SetHigh();
        __delay_ms(dash_duration);
        IO_RD3_SetLow();
        __delay_ms(dash_duration);
}
```

These functions define the letters S and O to be used in the SOS message:

```
void S() {
        dot();
        dot();
        dot();
        shortspace();
}
void O() {
        dash();
        dash();
        dash();
        shortspace();
}
```

The following is the main function, which defines the SOS message and its leading time space:

```
void main(void)
{
    SYSTEM_Initialize();
    IO_RD3_SetLow();
    while (1)
    {
        S(); O(); S();
        space();
    }
}
```

As you can see from the preceding code, RD3 is used as an output port for driving the 2N2222 transistor and therefore switching the LED on and off.

Bear in mind that the code uploaded to the GitHub repository contains many comments explaining its main parts.

The next section describes how can we test out a super-bright LED to see whether it works correctly, as well as testing the speed of the SOS message shown by the LED.

Testing the visual alarm

In this section, we will focus on how to test the super-bright LED, as well as how to test the speed of the SOS message shown by the LED.

You can test the super-bright LED to establish whether it works OK with a power supply, as shown in *Figure 6.10*:

Figure 6.10 – Connecting the super-bright LED to a battery set

The following are the steps for connecting everything according to *Figure 6.10*:

1. Connect the super-bright LED's anode (its longest leg) to a 1 k ohm resistor.

2. Connect the resistor to the battery set's positive terminal. The battery set should provide around 3 V (supplied by two AA batteries), which are enough for testing our super-bright LED.

3. Connect the battery set's negative terminal to the super-bright LED's cathode (its shortest leg).

After connecting everything as per the preceding steps, the LED should glow. If not, check the LED polarity, and whether the batteries have enough energy. Also check the connections. If the LED doesn't glow, it doesn't work and needs to be discarded.

If you compare the brightness of a regular LED against the super-bright LED that we used in this chapter, you will notice that the regular LED glows more uniformly and more **omnidirectional**, whereas the super-bright LED has a viewing angle of 35 degrees, meaning that it will be brighter from its top rather than from its sides. Try it! Observe very briefly both LEDs from the side. Remember that you shouldn't stare at the super-bright LED perpendicularly (top view), as it may hurt your eyes.

You can also test the speed of the SOS message. Change the value originally declared in the dot_duration variable in the preceding code for both the Blue Pill and Curiosity Nano. A smaller value will make the SOS message glow faster on the super-bright LED.

Summary

In this chapter, we learned what a super-bright LED is and how we can connect it to a microcontroller board. We also reviewed how to use the super-bright LED as a powerful visual alarm, since it glows much more intensely than conventional LEDs. We also summarized what Morse code is, how it is used worldwide, and how to show the SOS Morse code message, which is used in visual alarms as a distress message, by turning on and off the super-bright LED connected to the Blue Pill and Curiosity Nano microcontroller boards. Connecting a super-bright LED is not straightforward, since we will need to know how much current it will draw, because the microcontroller boards' output ports can handle a very limited amount of current, in the order of a few milliamps. This chapter will be beneficial for readers who would like to control an LED in other electronic projects. It also points out the importance of carefully calculating the amps that are drawn by a super-bright LED and using the right resistor or transistor to connect to it, so as to avoid damaging the microcontroller board.

The next chapter will focus on using a small microphone connected to a microcontroller board to detect two clapping sounds in a row to activate a process on a microcontroller board.

Further reading

- Choudhuri, K. B. R. (2017). *Learn Arduino Prototyping in 10 days*. Birmingham, UK: Packt Publishing Ltd.

- Gay, W. (2018). *Beginning STM32: Developing with FreeRTOS, libopencm3, and GCC*. New York, NY: Apress.

- Horowitz, P., Hill, W. (2015). *The Art of Electronics*. [3rd ed.] Cambridge University Press: New York, NY.

- Microchip (2019). *PIC16F15376 Curiosity Nano Hardware User Guide*. Microchip Technology, Inc. Available from: `http://ww1.microchip.com/downloads/en/DeviceDoc/50002900B.pdf`.

- 2N2222 (2013). P2N2222A transistor datasheet. Semiconductor Component Industries, LLC. Available from:

  `https://www.onsemi.com/pub/Collateral/P2N2222A-D.PDF`.

- LED (n.d.) BL-BJ33V4V-1 super-bright LED datasheet. Bright LED Electronics Corp. Available from: `http://www.maxim4u.com/download.php?id=1304920&pdfid=446ED6935162B290D3BC0AF8E0E068B8&file=0168\bl-bj33v4v-1_4138472.pdf`.

7
Creating a Clap Switch

When it comes to automating the home, one of the most desired features is often the ability to switch electronic devices on and off remotely (Colon, 2020). To address this, in this chapter – which covers the fifth project in this book – we will learn how to build a wireless electronic remote control to turn an LED on when two successive clapping sounds are detected using a simple microphone (**clap switch**). The next two successive clapping sounds detected will turn the LED off. The importance of having a switch clap remote control is that an electronic appliance can be turned on and off from anywhere in a room. This characteristic makes it of particular interest to the elderly or people with a motor disability.

This chapter will cover the following main topics:

- Connecting a microphone to a microcontroller board port
- Coding your clap switch sketch
- Coding a clap switch with two clapping sounds
- Coding a clap switch with a timer between claps
- Improving the project performance

After completing this chapter, you will be able to apply what you have learned to projects that need to read data from an analog source, process this data to convert it to digital data, and thus be able to use it to automate processes.

Technical requirements

The hardware components that will be needed to develop the clap switch are as follows:

- One breadboard
- One electret microphone FC-04 module
- Seven male-to-male jumper wires
- One LED
- One 220-ohm resistor
- A 5-volt power source

These components are very common, and there will be no problems in getting them easily. On the software side, you will require the Arduino IDE and the GitHub repository for this chapter: `https://github.com/PacktPublishing/DIY-Microcontroller-Projects-for-Hobbyists/tree/master/Chapter07`

The Code in Action video for this chapter can be found here: `https://bit.ly/3h2xjQu`

To supply the power source, you can use power adapters (*Figure 7.1*) that fit into the breadboard rails (these power adapters or breadboard power supplies are common and can be purchased at a low price in online stores), can be jumpered to provide 3.3 or 5 volts, and include an on/off button:

Figure 7.1 – Power supply

Additionally, you can use a 9-volt battery. Using these batteries is one of the most common options since the 9-volt voltage is appropriate to power up projects with microcontrollers such as the Blue Pill STM32.

Connecting a microphone to a microcontroller board port

In this section, we are going to learn about the hardware components needed to build a clap switch using the **STM32 Blue Pill** and the microphone FC-04 module.

However, before we begin connecting the components, we must understand the basics of a microphone.

Understanding the electret microphone module

A clap switch uses a microphone to sense the environment while waiting for an event that will trigger an action. In this section, we will understand how to apply this functionality in a project.

We will use a generic microphone module, which is a breakout board with an electret condenser microphone (as shown in *Figure 7.2*):

Figure 7.2 – Electret microphone board

Condenser microphones are composed of a diaphragm membrane on a plate, and both are conductors. Condensers are essentially capacitors formed of conductors and insulation between them. Therefore, when there is a smaller distance between these two conductors, a higher capacitance is obtained.

> **Important note**
> Capacitance is the property or capacity that an electronic component has to collect and store energy in the form of an electrical charge.

The microphone receives the sound, making the diaphragm vibrate. The vibration varies the distance between the conductors and changes its capacitance to produce a voltage charge, which in turn requires a lot of voltage to maintain, making traditional microphones ineffective for projects with microcontroller boards (such as the Blue Pill or the Curiosity Nano) in reducing power consumption. To address the issue of the high amount of power consumption, electret microphones were developed. Electret microphones use a unique polarized material charged during their manufacturing process, thus not requiring external voltage (Fox, 2020). With an understanding of microphone basics, we will now move on to the steps to connect a microphone to a microcontroller board.

Connecting the components

Now we are going to connect the electronic components to the breadboard, do the wiring, and finally connect everything to the STM32 Blue Pill:

1. In connecting the components, place the electret microphone, the resistor, the LED, and the STM32 Blue Pill on a breadboard with enough space to add the wiring layer, as shown in *Figure 7.3*. The hardware connections for this project are very straightforward:

Figure 7.3 – Components on the breadboard

2. Next, to power up the clap switch with an external power source, connect the 5-volt pin to the red rail on the breadboard and a ground pin to the blue track, as shown in the following photo (*Figure 7.4*):

Figure 7.4 – Connections to the power supply

3. Connect the ground (**GND**) pin of the sensor to the blue rail (the holes next to the blue lines) of the breadboard or a **GND** terminal of the SMT32 Blue Pill. Next, you need to connect the voltage (**VCC**) pin to the red rail (the holes next to the red lines) of the breadboard, or the **5V** bus of the Blue Pill, as shown in the following figure. The sensor generates an analog output, so it must be connected to an analog input on the Blue Pill card and connect the output pin of the sound sensor to pin **A0** of the Blue Pill, as shown in *Figure 7.5*:

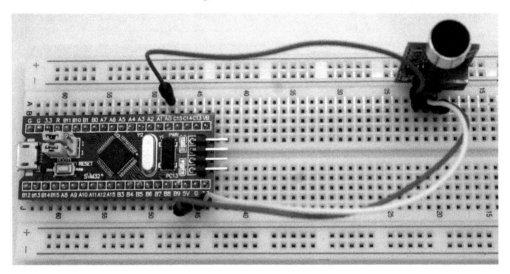

Figure 7.5 – Microphone connection

In this way, the acquisition of the analog signal that comes from the sensor will be achieved, and the microcontroller will convert it into digital.

4. To configure the LED, connect the cathode of the LED to a **GND** pin of the Blue Pill and the anode to pin **13** of the Blue Pill. The resistor must be between these two as this is a digital output pin (see *Figure 7.6*):

Figure 7.6 – LED setup

Finally, you need to use a power source such as batteries or the ST-LINK connected to the USB port of the computer to power up the board. The ST-LINK will also serve to upload the scripts to the microcontroller board. *Figure 7.7* summarizes all the hardware connections:

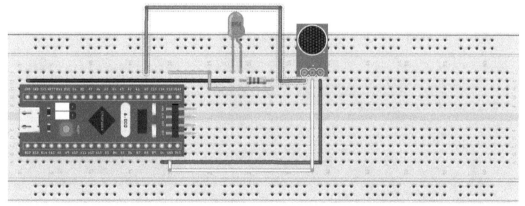

Figure 7.7 – Circuit for the microphone sensor connection

The previous figure shows all the connections between the STM32 Blue Pill and the electronic components. This figure summarizes the connection steps we just completed.

Figure 7.8 presents the schematics for this project:

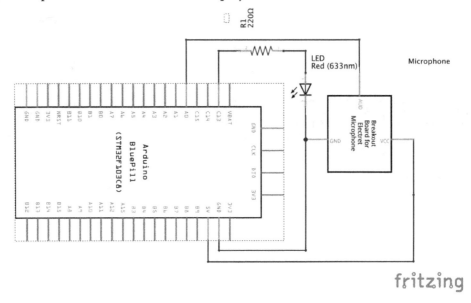

Figure 7.8 – Schematics for the microphone sensor connection

The schematics figure shows the electric diagram for the complete project. *Figure 7.9* shows how everything is connected in our **do it yourself** (**DIY**) clap switch:

Figure 7.9 – Clap switch device

Figure 7.9 shows how all the finished hardware connections will look.

Now, let's move on to the next section, which will detail the essential parts of the C code to complete the functionality of the clap switch.

Coding your clap switch sketch

In this section, we will develop the program to identify a clap sound from a microphone. This sound will turn an LED on and off. Let's get started:

1. As a first step, we need to define which pins of the Blue Pill card pins will be used for input and output. Then, we need to assign the sound threshold level for the microphone to detect the sound; this value is in the range 0-1023. We are using a value of 300, so the sound captured by the microphone is loud enough to identify a clap and not any background noise (we will show how to select an appropriate threshold in the *Improving the project performance* section).

    As can be seen in the following code snippet, the analog reading pin will be 0 (labeled **A0** on the Blue Pill), and the digital output pin will be PC13 (labeled **C13** on the Blue Pill):

    ```
    const int MicAnalogPin = 0;
    const int LedDigitalPin = PC13;
    const int ClapThreshold = 300;
    ```

    Constant value variables are being used with the const keyword to assign its value. Another way to do it is with a preprocessor macro using #define, as shown in the following code snippet:

    ```
    #define MicAnalogPin 0
    #define LedDigitalPin PC13
    #define ClapThreshold 300
    ```

    The difference between both forms is that with #define, the compiler replaces all occurrences with the values before compilation to avoid memory usage in the microcontroller and will always have a global scope. In contrast, const is a constant value variable stored in the microcontroller memory and has a limited scope.

    Based on best programming practices, the recommended option for type safety is the use of constant variables, given that the #define directive substitutes the macro value disregarding the scope, which can lead to data type problems. In contrast, const will always be of the same data type that was defined in its declaration.

2. Next, in the `setup()` part, we need to start the serial data transmission and assign the speed of the transfer (`9600` bps as a standard value):

```
void setup() {
    Serial.begin(9600);
}
```

3. We also need to configure which pins the microcontroller will use as input and output. These values were previously defined in the constants. In the following code, the assignment is made to the microcontroller card:

```
void setup() {
    Serial.begin(9600);
    pinMode(LedDigitalPin, OUTPUT);
    pinMode(MicAnalogPin, INPUT);
}
```

4. Now comes the `loop()` part in the sketch. It contains two main parts: *the reading of the analog input pin* and *the clap detection to turn on the LED*.

5. The `analogRead()` function reads the value of the previously defined input pin:

```
void loop() {
    int SoundValue = analogRead(MicAnalogPin);
    Serial.print("Sound value: ");
    Serial.println(SoundValue);
}
```

6. Now we have the read value from the microphone in the `SoundValue` variable; the next step will be to compare its value with the defined threshold. If the sound input by the microphone is higher than the threshold, the LED will light up, and a 1-second pause will occur, in which you can observe that the LED is lit. If the detected sound does not exceed the threshold value, the LED will be instructed to remain off. In both cases, the serial console will display the status of the running script:

```
void loop() {
    int SoundValue = analogRead(MicAnalogPin);
    Serial.print("Sound value: ");
    Serial.println(SoundValue);
    if (SoundValue > ClapThreshold) {
        Serial.println("A clap was detected");
```

```
        digitalWrite(LedDigitalPin, HIGH);
        delay(1000);
    } else {
        Serial.println("No clap detected");
        digitalWrite(LedDigitalPin, LOW);
    }
}
```

Now we have the complete code for the first sketch to detect the sound of a clap and turn on a LED. Next, we can see the complete sketch, which is available in the Chapter7/ clap_switch folder in the GitHub repository.

Now that the sketch is complete, you can upload it to the Blue Pill board. To test that our project works, just give a clap to see how the LED turns on. Do not forget that the breadboard must be powered, either by batteries or connected to the computer.

So far, we have learned how to read an analog value (sound) from a microphone with a microcontroller. During the main loop, the device keeps listening to the microphone to detect whether any sound (we will assume the detected sound is a clap) is louder than the defined threshold; if so, an LED lights up. If not, the microcontroller will keep the LED off.

Next, we are going to modify the sketch to indicate to our clap switch to wait for two claps before turning on the LED. For better organization, we will create a copy of the code to compare with the original sketch or in case we need to revert to the previous version when experiencing an issue with the new code.

Coding a clap switch with two clapping sounds

In this section, we will modify our program to identify two clapping sounds from the microphone. This will allow us to be more precise before activating the remote control:

1. Once we define the constants, we define two variables: the integer type, to count the number of claps, and the Boolean type, to know the status of the LED (on or off). For a better reading, we have highlighted the changes to the original variables' declaration of the sketch:

    ```
    const int MicAnalogPin = 0;
    const int LedDigitalPin = PC13;
    const int ClapThreshold = 300;
    int ClapNumber = 0;
    bool LedState = false;
    ```

As you can see, the clap count is set to 0, and the LED status to `false` (off), which means that `true` will be on.

2. We will leave the `setup()` section unchanged and continue to the `loop()` section, where we have the most important logic changes. We will modify the instructions that are within the conditional sentence that verifies whether the sound registered by the microphone is louder than the defined threshold:

```
void loop() {
    int SoundValue = analogRead(MicAnalogPin);
    if (SoundValue > ClapThreshold) {
      ClapNumber++
      delay(500);
    }
}
```

This conditional block does not require an `else` sentence. Also, the instruction to turn on the LED has been removed. If a clap is detected, the clap counter will be incremented by 1 and will wait for a half-second pause before the Blue Pill re-senses the microphone.

3. We need to know whether the total number of claps is 2, so we must write another condition in our sketch to ask whether the `ClapNumber` variable already has the two claps registered; if so, it must turn on the LED:

```
void loop() {
    int SoundValue = analogRead(MicAnalogPin);
    if (SoundValue > ClapThreshold) {
      ClapNumber++;
      delay(500);
    }
    if (ClapNumber == 2) {
      digitalWrite(LedDigitalPin, HIGH);
    }
}
```

Now we are ready to test the new sketch on our device, compile it, and load it. We can test the code with two consecutive claps, with a short pause between them (remember the delay that we programmed).

4. As you can perceive, the LED turns on, but it does not turn off. For turning off the LED again, we are going to program the functionality to turn off when the user gives two more claps. For this, we will require the other variable that we define: `LedState`. In it, we will store whether the LED is on or off:

```
void loop() {
    int SoundValue = analogRead(MicAnalogPin);
    if (SoundValue > ClapThreshold) {
      ClapNumber++;
      delay(500);
    }
    if (ClapNumber == 2) {
      if (LedState) {
        digitalWrite(LedDigitalPin, HIGH);
      } else {
        digitalWrite(LedDigitalPin, LOW);
      }
      ClapNumber = 0;
      LedState = !LedState;
    }
}
```

When the microcontroller detects two claps in a row, the first thing it will do is check the state of the variable that stores the LED state using the next conditional sentence, `if (LedState)`. As it is a Boolean type variable, it can be analyzed directly into the `if` statement without the need for additional operators. If the variable has a TRUE value, it triggers the `if` block, and if it is FALSE, it executes the `else` block. So, the preceding code turns on the LED if the variable is TRUE and turns it off if it is FALSE. Finally, within the same condition of two claps detected, the `ClapNumber` variable is reset to 0 claps to restart the counter and wait for two new claps. The value of the `LedState` variable is changed utilizing a negation operator, inverting the value of the Boolean variable; if the state was ON, the command would turn it into OFF and vice versa.

With this previous step, our project for detecting two continuous claps and turning on an LED is complete. The full sketch is shown as follows and is also available in the `Chapter7/double_clap_switch` folder in the GitHub repository. With the complete sketch for detecting two continuous claps, we can already test it with its new functionality. In the beginning, the LED will be off. When you clap twice, with a short pause between them, it will light up; clapping twice again will turn it off.

Finally, we are going to add a timer to our sketch to indicate to our clap switch to wait only 30 seconds between the two claps before turning on the LED. The same as this section, we will create a copy of the code and work with the new code.

Coding a clap switch with a timer between claps

Now, we will add a timer to limit the waiting timeframe between the first and second claps:

1. Define two new variables, both of the unsigned long type, to store the time of each clap. The changes to the previous sketch are highlighted:

```
const int MicAnalogPin = 0;
const int LedDigitalPin = PC13;
const int ClapThreshold = 300;
int ClapNumber = 0;
bool LedState = false;
unsigned long FirstClapEvent = 0;
unsigned long SecondClapEvent = 0;
```

These two highlighted variables will store the milliseconds when a clap is detected: one for the first time and another for the second time.

2. The setup() section remains unchanged, and we will continue to the loop() section to introduce the timer changes. We will add a conditional inside the conditional sentence for detecting a clap:

```
void loop() {
    int SoundValue = analogRead(MicAnalogPin);
    if (SoundValue > ClapThreshold) {
        ClapNumber++
        delay(500);
        FirstClapEvent = (ClapNumber == 1) ? millis() :
            FirstClapEvent;
    }
}
```

If a clap is detected, we verify whether it is the first clap; in this case, we assign the FirstClapEvent variable value to the millis() function. This function returns the milliseconds since the STM32 Blue Pill executed the sketch. Otherwise, the variable value remains the same.

3. After we identify the two claps, we will assign `millis()` to the
   `SecondClapEvent` variable. Now, we need to write another condition in our
   script to ask whether the time between both claps is less than 30 seconds:

```
void loop() {
   int SoundValue = analogRead(MicAnalogPin);
   if (SoundValue > ClapThreshold) {
     ClapNumber++;
     delay(500);
     FirstClapEvent = (ClapNumber == 1) ? millis() :
         FirstClapEvent;
   }
   if (ClapNumber == 2) {
     SecondClapEvent = millis();
     if (SecondClapEvent - FirstClapEvent < 30000) {
       if (LedState) {
         digitalWrite(LedDigitalPin, HIGH);
       } else {
         digitalWrite(LedDigitalPin, LOW);
       }
       ClapNumber = 0;
       LedState = !LedState;
       FirstClapEvent = 0;
       SecondClapEvent = 0;
     }
   }
}
```

Now we can test the code with two consecutive claps, with a 30-second timeframe
between them. The full sketch is shown as follows and is also available in the GitHub
repository folder: `Chapter7/double_clap_switch_timer`.

Now that we have completed the sketch with all its functionality, just load and execute the
sketch to test and validate the functionality.

In the next section, we will learn how to test the system and provide a useful tip to
improve the work with analog data.

Improving the project performance

I mentioned earlier that we were going to talk about the threshold value. In general, solutions for projects with analog read set the threshold value to 200, but as you remember, we are using a value of 300 to ensure we are reading a clap, not background noise.

If you want to know more precisely the value of your clap's sound, then you can use the example script that the IDE provides us. To access this sketch, we must go to the **File** menu, then to **Examples**, and choose **Basics**. In that section, we will find the **AnalogReadSerial** sketch, as shown in *Figure 7.10*:

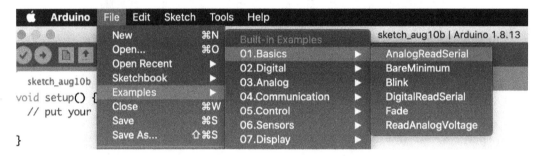

Figure 7.10 – Analog serial reading example

Selecting that menu option will open the `AnalogReadSerial.ino` sketch:

```
void setup() {
   Serial.begin(9600);
}


void loop() {
   int sensorValue = analogRead(0);
   Serial.println(sensorValue);
   delay(1);
}
```

The previous code introduces an instruction that we have not used so far in this project, which is `Serial.println()`. This code displays the data in ASCII format in a console called the serial port so that it is understandable to people and it includes a line break.

By loading and executing the script, we can open the serial monitor of our Arduino IDE to see the values that our microphone is generating (in the range 0-1023). You just need to try some test claps to find out the value of the sound of your clap and use that as a more personalized threshold.

The previous code blocks sense a microphone to detect two clap sounds and turn on and off an LED. The system waits for a 30-second timeframe between each clap sound before turning on or off the LED.

Congratulations, you have now built a wireless electronic remote control!

Summary

So, what have we learned in this project? Firstly, we learned how to connect the microphone module, LED, and resistor to the STM32 Blue Pill microcontroller board, which would be controlled by the STM32 microcontroller. We then wrote a piece of code to read an analog value and analyze it in our microcontroller. Subsequently, we sent a digital signal to turn the LED on or off depending on the embedded rules in our STM32. Lastly, we physically tested the device to understand the real-life operation.

This project gives us the skills to begin to create a remote control to automate home appliances and use them according to our needs. For example, you can add a relay module and connect it to a lamp so you can turn it on and off from the comfort of where you are without the need to reach the lamp switch.

In the next chapter, we will learn how to use the serial monitor feature to analyze the outputs that our sketch generates while it is running. We will do so by building a gas sensor project with the help of the STM32 microcontroller.

Further reading

- Colon, A. (2020). *The Best Smart Home Devices for 2020*. PCMAG: `https://www.pcmag.com/news/the-best-smart-home-devices-for-2020`
- Fox, A. (2020). *The Complete Guide to Electret Condenser Microphones*. My New Microphone: `https://mynewmicrophone.com/the-complete-guide-to-electret-condenser-microphones/`

8
Gas Sensor

An indoor environment with good air quality is essential to guarantee a healthy environment (Marques and Pitarma, 2017). The **MQ-2 gas sensor** can be an excellent way to measure the quality parameters of indoor air or as an early fire detection system. In this chapter, you will learn how to build a practical system for detecting gases in the environment (which we will call a **gas sensor**) and connect the MQ-2 gas sensor to a **Blue Pill microcontroller card**.

The following main topics will be covered in this chapter:

- Introducing the MQ-2 gas sensor
- Connecting a gas sensor to the STM32 microcontroller board
- Writing a program to read the gas concentration over the sensor board
- Testing the system

At the end of this chapter, you will know about the operation of an MQ-2 gas sensor, and you will be able to connect it correctly to the STM32 microcontroller card and view the data obtained from the sensor. You will be able to apply what you have learned in projects that require the use of sensors to detect substances such as flammable gases or alcohol or measure air quality.

Technical requirements

The hardware components that will be needed to develop the gas sensor are as follows:

- One solderless breadboard
- One Blue Pill board
- ST-Link/V2
- One MQ-2 breakout module
- Seven male-to-male jumper wires
- One LED 8x8 matrix
- One 7219 breakout board
- A 5 V power source

These components are widespread, and there will be no problems in getting them easily. On the software side, you will require the Arduino IDE and the GitHub repository for this chapter: `https://github.com/PacktPublishing/DIY-Microcontroller-Projects-for-Hobbyists/tree/master/Chapter08`

The Code in Action video for this chapter can be found here: `https://bit.ly/2UpGDGs`

Let's first start by describing the characteristics of the MQ-2 gas sensor.

Introducing the MQ-2 gas sensor

In this section, we will get to know the details of the main hardware component to build our gas sensor: the **MQ-2 sensor**. This sensor is recommended to detect LPG, propane, alcohol, and smoke, mainly with concentrations between 300 and 10,000 **parts per million (ppm)**. So, we can say that it is a sensor to detect smoke and flammable gases.

Concentration refers to the amount of gas in the air and is measured in ppm. That is, if you have 2,000 ppm of LPG, it means that in a million gas molecules, only 2,000 ppm would be LPG and 998,000 ppm other gases.

The MQ-2 gas sensor is an electrochemical sensor that varies its resistance when exposed to certain gasses. It includes a small heater to raise the circuit's internal temperature, which provides the necessary conditions for the detection of substances. With the 5 V connection on the pins, the sensor is kept warm enough to function correctly.

> **Important note**
> The sensor can get very hot, so it should not be touched during operation.

The MQ series gas sensors are analog, making them easy to implement with any microcontroller card, such as the STM32 Blue Pill. It is very common to find the MQ-2 sensor in breakout modules, which facilitates connection and use since it will only be necessary to power it up and start reading its data. These breakout modules have a **digital output** (**DO**) that we can interpret as the presence (*LOW*) or absence (*HIGH*) of any gas detected by the sensor. *Figure 8.1* shows the MQ-2 gas sensor with a breakout board:

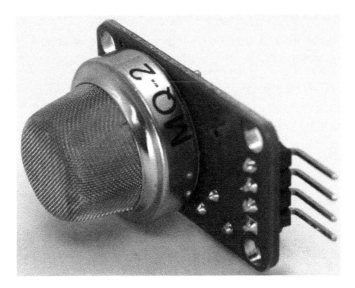

Figure 8.1 – MQ-2 gas sensor with a breakout board

In the next section, we will learn how to connect the MQ-2 sensor to our solderless breadboard to obtain its reading data through digital and analog means.

Connecting a gas sensor to the STM32 microcontroller board

In this section, we will build a gas sensor device utilizing the **STM32 Blue Pill** microcontroller board and a gas sensor module using the hardware components listed in the *Technical requirements* section. The gas sensor breakout board connects to the STM32 Blue Pill with four pins:

- **Analog output** (**AO**): This pin generates an analog signal and must be connected to an analog input of the microcontroller.

- **DO**: This pin generates a digital signal and must be connected to a digital input of the microcontroller.

- **VCC**: Pin to supply power to the sensor (5 V).
- **GND**: Ground connection.

For this project, you will learn how to interface the MQ-2 module with the STM32 board to acquire data in a digital and analog way. Let's start with the digital option.

Interfacing for digital reading

Now we are going to connect the electronic components to the breadboard, do the wiring, and finally connect everything to the STM32 Blue Pill:

1. In connecting the components, place the sensor module and the STM32 Blue Pill on a solderless breadboard with enough space to add the wiring layer, as shown in *Figure 8.2*. The hardware connections for this project are exceptionally effortless:

Figure 8.2 – Components on the breadboard

2. Next, power up the Blue Pill with an external power source. Connect the **5 V pin** to the red rail on the breadboard and a **G pin** to the blue track, as shown in *Figure 8.3*:

Figure 8.3 – Connections to the power supply

3. Connect the **GND pin** of the MQ-2 sensor to a GND terminal of the SMT32 Blue Pill. Next, you need to connect the **VCC pin** to the 5 V bus of the Blue Pill, as shown in the following figure. In this section, we will read the DO, so it must be connected to a digital input on the Blue Pill card. Connect the DO of the MQ-2 sensor to pin B12 of the Blue Pill, as shown in *Figure 8.4*:

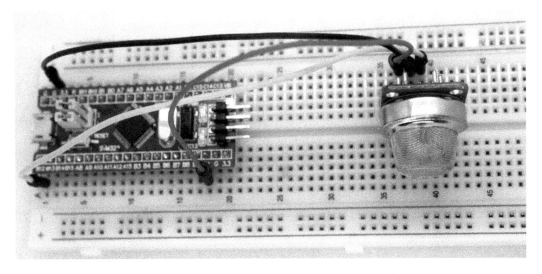

Figure 8.4 – MQ-2 sensor connection for digital reading

Finally, you need to use a power source such as batteries to power up the board. *Figure 8.5* summarizes all the hardware connections:

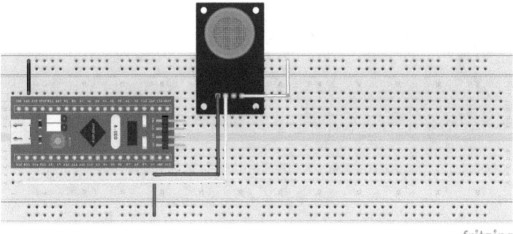

Figure 8.5 – Circuit for the MQ-2 sensor connection for digital reading

The preceding figure shows all the connections between the STM32 Blue Pill and the electronic parts. *Figure 8.6* presents the schematics for this project:

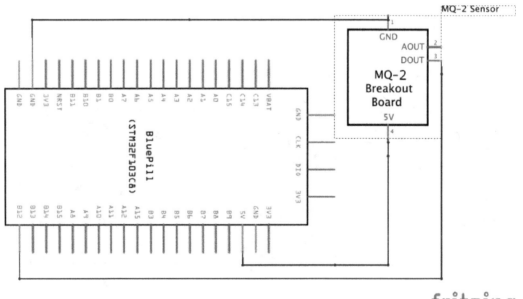

Figure 8.6 – Schematics for the MQ-2 sensor connection for digital reading

Figure 8.7 shows how everything must be connected in our DIY gas sensor device:

Figure 8.7 – Gas sensor device for digital reading

In this subsection, we learned how to connect the electronics to create our gas sensor device with digital reading. Next, we will see how to connect it so that the reading is analog.

Interfacing for analog reading

Only one step will be necessary to change how our hardware device reads data from the sensor to be an analog reading instead of a digital one:

1. Disconnect the jumper wire from the DO pin and connect it to the AO pin of the MQ-2 sensor. Also, instead of connecting to pin B12, connect to pin AO of the Blue Pill, as shown in *Figure 8.8*:

Figure 8.8 – MQ-2 sensor connection for analog reading

Figure 8.9 summarizes all the hardware connections:

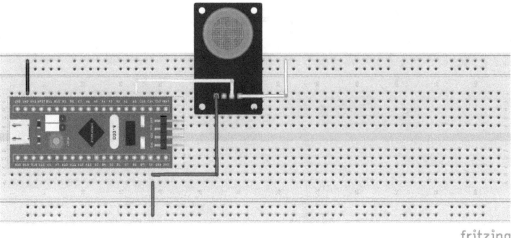

Figure 8.9 – Circuit for the MQ-2 sensor connection for analog reading

Figure 8.10 presents the schematics for the analog reading device:

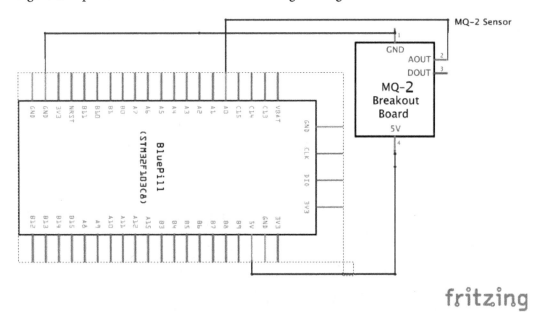

Figure 8.10 – Schematics for the MQ-2 sensor connection for analog reading

Let's recap. In this section, we learned how to connect the hardware components to create our gas sensor device. You learned how to connect the MQ-2 sensor to the STM32 Blue Pill microcontroller board to obtain its data in two ways: digitally and in analog form.

In the next section, we will create the C code that obtains the MQ-2 sensor data from the STM32 Blue Pill microcontroller.

Writing a program to read the gas concentration over the sensor board

In this section, we will learn how to code a program to read data from our gas sensor and show it on the serial monitor if gas is present in the environment.

As in the previous section, we'll first learn how to read data digitally and also in analog form.

Coding for digital reading

Let's start writing the code:

1. Define which pin of the STM32 Blue Pill microcontroller will be used as input for reading the data from the sensor. Here's the code that shows how to do that:

    ```
    const int sensorPin = PB12;
    boolean sensorValue = true;
    ```

    The selected pin was PB12 (labeled B12 on the Blue Pill board). A Boolean variable was declared and initialized to true. This variable will be used for storing the sensor data.

2. Next, in the setup() part, we need to start the serial data transmission and assign the speed of the transfer (9600 bps as a standard value):

    ```
    void setup() {
      Serial.begin(9600);
    }
    ```

3. Indicate to the microcontroller the type of pin assigned to `PB12`:

```
void setup() {
    Serial.begin(9600);
    pinMode(sensorPin, INPUT);
}
```

4. Now comes `loop()` with the rest of the sketch. The first lines read the input pin's data sensor and display its value in the serial console:

```
void loop() {
    sensorValue = digitalRead(sensorPin);
    Serial.print("Sensor value: ");
    Serial.println(sensorValue);
    if (sensorValue) {
        Serial.println("No gas present");
        delay(1000);
    } else {
        Serial.println("Gas presence detected");
        delay(1000);
    }
}
```

The value read from the sensor could be TRUE or FALSE; remember, we are reading a digital value. If the value is TRUE, then gas is not present in the environment; otherwise, gas was detected. This behavior occurs because the MQ-2 sensor has a negated output; the module's LED must also light up in this state since it is internally with a 5 V resistance. When there is no presence of gas, the LED turns off, and the output is logic 1 (5 V).

The code for digital reading is now complete. You can find the complete sketch available in the `Chapter8/gas_digital` folder in the GitHub repository.

Now we have the complete code for reading the DO of the MQ-2 sensor. You can upload it to the STM32 microcontroller. You can now see, in the **serial monitor**, the sensor readings as shown in *Figure 8.11*. The most normal thing is for the reading to indicate no presence of any gas:

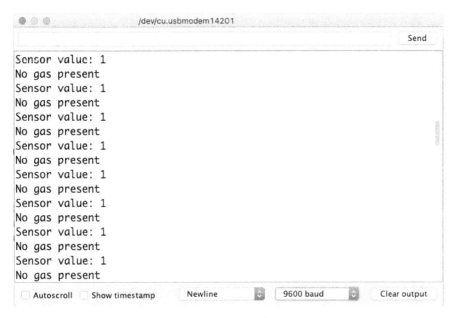

Figure 8.11 – Serial monitor reading the DO of the sensor with no gas presence

Now, **being very careful about fire safety**, bring a lit match to the sensor and put it out when close to the sensor to generate smoke. The serial monitor will change as soon as the smoke impregnates the sensor (as shown in *Figure 8.12*):

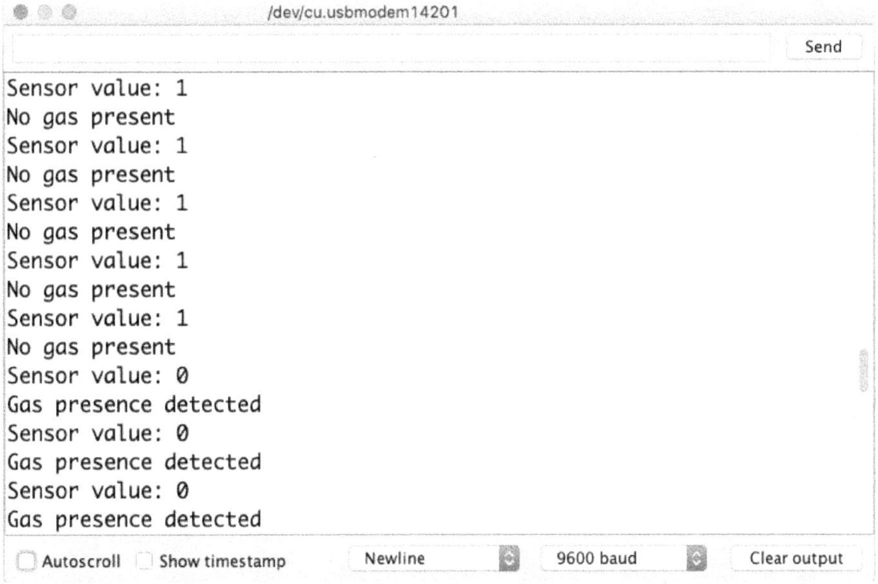

Figure 8.12 – Serial monitor reading the DO of the sensor with the presence of gas

As we can see, it is like reading any digital input. The sensitivity of the sensor is configured through the variable resistance included in the breakout module. Turning to the right becomes more sensitive, and we need less gas present to activate the output. In the same way, if we turn it to the left, a more significant presence of gas will be needed to activate the output.

So far, we have learned how to read the gas sensor in digital form. In the following subsection, we are going to obtain its reading from the AO.

Coding for analog reading

When using the AO, different levels of gas presence are obtained. The module has a heating chamber in which the gas enters. This gas will continue to be detected until the chamber is empty. The sensor's voltage output will be proportional to the gas concentration in the chamber.

In Short, the higher the gas concentration, the higher the voltage output, and the lower the gas concentration, the lower the voltage output.

Let's get started with the code:

1. Create a copy of the Chapter8/gas_digital project and change the name to Chapter8/gas_analog. Remember to rename the folder and the INO file.

2. Change the sensor pin to 0 (labeled A0 on the Blue Pill), remove the Boolean variable, and assign a threshold level for the sensor readings. We will use a value of 800, to be sure the sensor has gas in its chamber:

```
const int sensorPin = 0;
const int gasThreshold = 800;
```

3. Keep setup() without modifications:

```
void setup() {
  Serial.begin(9600);
  pinMode(sensorPin, INPUT);
}
```

4. The code in loop() will be using the same logic but with a few changes:

```
void loop() {
  int sensorValue = analogRead(sensorPin);
  Serial.print("Sensor value: ");
```

```
Serial.println(sensorValue);
if (sensorValue > gasThreshold) {
  Serial.println("Gas presence detected");
} else {
  Serial.println("No gas present");
}
delay(1000);
}
```

5. To read the sensor value, we use the `analogRead()` function. The value read is stored in the `sensorValue` variable; the next step will be to compare its value with the threshold. If the sensor value is higher than the threshold, this means gas was detected.

Now that our sketch is complete, upload it to the Blue Pill board. To test that our project works, just like the digital reading version, bring a lit match to the sensor and put it out when close to the sensor to generate smoke. Please, do not forget to be very careful about fire safety. *Figure 8.13* shows the serial monitor when smoke starts to be detected:

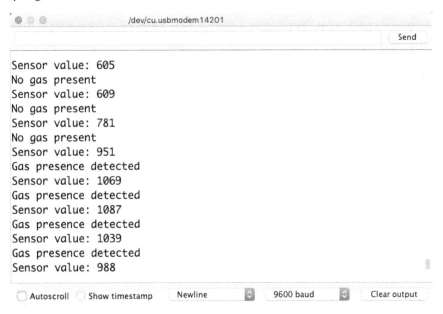

Figure 8.13 – Serial monitor reading the AO of the sensor with gas present

This section helped us to learn how to create code in C to read the data from the MQ-2 sensor to know if there is a concentration of gas or smoke in the environment. In addition, the skills to read the sensor value in both analog and digital form were acquired. In the next section, we will create a simple way of knowing directly in the hardware device if there is gas or smoke concentration without seeing the serial monitor on a computer.

Testing the system

In this last section of the chapter, we will connect an **8x8 LED matrix** to display an alert if the sensor detects the presence of gas in the environment.

An LED matrix is a set of LEDs grouped into rows and columns. By turning on these LEDs, you can create graphics or text, which are widely used for billboards and traffic signs.

There is an electronic component for small-scale projects called an 8x8 LED matrix. It is composed of 64 LEDs arranged in eight rows and eight columns (see *Figure 8.14*):

Figure 8.14 – LED matrix 8x8

As you can see in the previous figure, the 8x8 LED matrix has pins to control the rows and columns, so it is impossible to control each LED independently.

This limitation implies having to use 16 digital signals and refreshes the image or text continuously. Therefore, the integrated MAX7219 and MAX7221 circuits have been created to facilitate this task; the circuits are almost identical and interchangeable using the same code.

In addition to these integrated circuits, breakout modules have been created integrating the 8x8 LED matrix and the MAX7219 circuit, in addition to having output connectors to put several modules in a cascade. *Figure 8.15* shows the 8x8 LED matrix breakout module:

Figure 8.15 – LED matrix 8x8 breakout module

The input pins of the module are as follows:

- **VCC**: Module power supply
- **GND**: Ground connection
- **DIN**: Serial data input
- **CS**: Chip select input
- **CLK**: Serial clock input

The output pins are almost identical, only instead of **DIN** there is **DOUT**, which will allow cascading with other modules, but we will not learn about this functionality in this chapter.

Figure 8.16 shows how to connect the MAX7219 8x8 LED matrix module to our STM32 Blue Pill board:

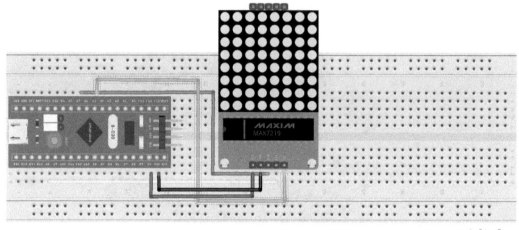

Figure 8.16 – LED matrix 8x8 breakout module interfacing to the STM32 Blue Pill

Now it is time to create the code to display an alert on our LED matrix. We will update the `Chapter8/gas_digital` sketch. Let's start coding!

1. To make the process easier, we will use a library called `LedControlMS` that facilitates the use of the 8x8 LED matrix module. To start with the installation, download the library from our GitHub: `Chapter8/library`.

2. To install, go to the **Sketch** menu | **Include Library** | **Add .ZIP Library…** (see *Figure 8.17*) and select the downloaded file, and it is ready to be used:

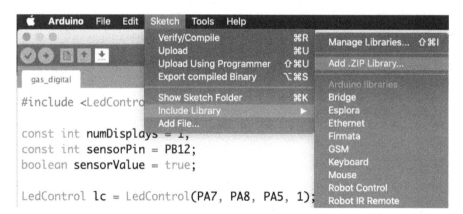

Figure 8.17 – Adding the LedControlMS library

3. In our script, we are going to add the library:

```
#include "LedControlMS.h";
```

4. We must indicate the number of display modules that we are using; in our case, it is one. We will initialize the library, pointing out the pins of the STM32 Blue Pill board to which the module will be connected, as well as the variable with the number of modules:

```
const int numDisplays = 1;
const int sensorPin = PB12;
boolean sensorValue = true;
LedControl lc = LedControl(7, 8, 5, numDisplays);
```

5. By default, the matrix is in power-saving mode, so it is necessary to wake it up. If it were more than one module, a loop would be required, but in this case, it is only one, so we will do it directly:

```
void setup() {
  Serial.begin(9600);
  pinMode(sensorPin, INPUT);
  lc.shutdown(0,false);
  lc.setIntensity(0,8);
  lc.clearDisplay(0);
}
```

The previous three lines do the following: activate the matrix, adjust the brightness, and clean all the LEDs. The value of 0 in the code refers to the first array of a possible set of interconnected arrays.

6. Finally, we write the character we want to show. We will use the `writeString()` function in the `else` statement to indicate in the LED matrix that there is gas; we will show a letter A representing an alert:

```
void loop() {
  sensorValue = digitalRead(sensorPin);
  Serial.print("Sensor value: ");
  Serial.println(sensorValue);
  if (sensorValue) {
    Serial.println("No gas present");
    delay(1000);
```

```
    } else {
      Serial.println("Gas presence detected");
      lc.writeString(0, "A");
      delay(1000);
    }
  }
```

We are ready to upload our script to the microcontroller and test that the system works. As in the previous section, to test it, bring a lit match to the sensor and put it out when close to the sensor to generate smoke. Again, do not forget to be very careful about fire safety. *Figure 8.18* shows the complete gas sensing device, including the sensor and LED matrix module connected to the STM32 microcontroller:

Figure 8.18 – Gas sensor device

Until now, in this section, we have learned how to handle an 8x8 LED matrix and use it to have a visual alert on our gas sensor device.

In this chapter, we learned how to read to code programs to read a gas sensor in a digital and analog way. This has allowed us to reinforce our knowledge of data acquisition from sensors in different forms of outputs. This knowledge will empower us to create more complex embedded systems, such as automating homes using sensors in the environment.

Summary

We had so much to learn in this chapter! First, we learned how to connect the MQ-2 gas sensor to the STM32 Blue Pill microcontroller board, both digitally and with an AO reading. We then wrote two pieces of code to read digital and analog sensor values. Last, we tested the device to understand its operation, displaying the sensor data in the serial console.

This project gave us the skills to read different kinds of sensor data to use this knowledge according to our needs. For instance, you can display some sensors in a room to monitor the environment in real-time.

In the next chapter, we will enter the fascinating world of the so-called Internet of Things. With the knowledge that we will acquire, we will create projects that connect to the internet and access our information remotely.

Further reading

Marques G. & Pitarma R. (2017). *Monitoring Health Factors in Indoor Living Environments Using Internet of Things.* In: Rocha Á., Correia A., Adeli H., Reis L., & Costanzo S. (eds) Recent Advances in Information Systems and Technologies. WorldCIST 2017. Advances in Intelligent Systems and Computing, vol. 570. Springer, Cham. `https://doi.org/10.1007/978-3-319-56538-5_79`

9
IoT Temperature-Logging System

In recent years, the use of the internet has increased. This same increase has allowed the internet to evolve. Now we speak of *things* connected to this network; devices for everyday use that were not originally designed to have connectivity. This evolution has created the concept of the **Internet of Things (IoT)**, which is defined by Morgan in Forbes (2014) as the *"interconnection to the internet of commonly used devices that can complete tasks in an automated way."*

The IoT is present in practically all fields of daily life, from health to education, known as the **Internet of Medical Things (IoMT)** and the **Internet of Educational Things (IoET)**, respectively.

In this chapter, you will be introduced to the world of creating IoT applications with a temperature logging application for an STM32 Blue Pill board using an **ESP8266 Wi-Fi module**. With this knowledge, you will be able to build projects that can connect to the internet and present their data from sources such as sensors remotely.

In this chapter, we will cover the following main topics:

- Connecting a temperature sensor to the Blue Pill board
- Coding a temperature reading system
- Learning to connect the ESP8266 module
- Coding a program to send the sensed temperature to the internet
- Connecting the STM32 Blue Pill board to the internet

By the end of this chapter, you will be able to understand the operation of one of the most popular Wi-Fi modules for creating IoT applications, the ESP8266, and also be familiar with how to connect the STM32 microcontroller card to the internet and send the data obtained from the temperature sensor.

Technical requirements

The hardware components that will be needed to develop the temperature-logging system are as follows:

- 1 solderless breadboard.
- 1 Blue Pill STM32 microcontroller board.
- 1 ST-Link/V2 electronic interface needed for uploading the compiled code to the Blue Pill board. Bear in mind that the ST-Link/V2 requires 4 female to female jumper wires.
- 1 DS18B20 temperature sensor module.
- 1 ESP8266 Wi-Fi module.
- 1 FTDI adapter board.
- 1 LED.
- 1 220 ohm resistor.
- 7 male to male jumper wires.
- 5 female to female jumper wires.
- A 5 V power source.

As usual, these components are very common, and there will be no problems in obtaining them. On the software side, you will require the Arduino IDE and the GitHub repository for this chapter: `https://github.com/PacktPublishing/DIY-Microcontroller-Projects-for-Hobbyists/tree/master/Chapter09`

The Code in Action video for this chapter can be found here: `https://bit.ly/3vSwPSu`

The following section presents an introduction to the temperature sensor module and its main features.

Connecting a temperature sensor to the Blue Pill board

In this section, we are going to learn the hardware components needed to build a temperature-logging sensor using the **STM32 Blue Pill** and a temperature module.

To build an electronic device that measures temperature, you will need a sensor that monitors the environment and records temperature data. A microcontroller card is also necessary to be able to read the data from the sensor and to be able to display the information to users. We will begin by having a look at the temperature sensor module.

Introducing the DS18B20 temperature sensor module

Let's get to know the main hardware component's details to build the temperature log: the **DS18B20 sensor**. It is a digital temperature sensor that can measure air temperature, liquids (using a waterproof version), and soil.

> **Important note**
>
> The DS18B20 temperature sensor has a unique 64-bit serial code, allowing multiple sensors to be connected using just one digital pin (1-wire protocol) from the STM32 microcontroller card.

We will use a generic breakout module that already includes a voltage LED and the required 4.7 kΩ pull-up resistor (as shown in *Figure 9.1*):

Figure 9.1 – DS18B20 digital temperature sensor breakout board

It is a one-wire sensor, which means the sensor requires only one pin port for communication with the controller. Temperature is measured in degrees Celsius from –55 °C to +125 °C with an accuracy of +/-0.5 °C (between -10 °C and 85 °C). The main advantage of using this sensor instead of a thermally sensitive resistor (thermistor) is that we receive from the sensor a stream of bits on a digital pin instead of receiving voltage on an analog pin.

Now that we know about the temperature sensor, let's move on to the following subsection to connect it to the STM32 Blue Pill on the solderless breadboard.

Connecting the components

We will connect the electronic components to the solderless breadboard, do the wiring, and finally connect everything to the STM32 Blue Pill. The following are the steps to be performed:

1. Place the temperature sensor and the STM32 Blue Pill on a solderless breadboard with enough space to add the wiring layer, as shown in *Figure 9.2*:

Figure 9.2 – Components on the breadboard

2. Next, we will power up the temperature-logging system with an external power source. To do this, connect the STM32 Blue Pill's 5 V pin to the red rail on the solderless breadboard and a ground pin to the blue track, as shown in the following photo (*Figure 9.3*):

Figure 9.3 – Connections to the power supply

3. Connect the ground (GND) pin of the sensor to the blue rail of the solderless breadboard or a GND terminal of the STM32 Blue Pill. Next, you need to connect the voltage (VCC) pin to the red rail of the solderless breadboard, or the 5 V bus of the STM32 Blue Pill, as shown in the following photo. The temperature sensor generates a digital output, so it must be connected to a digital input on the STM32 Blue Pill card. Connect the signal pin (S) of the temperature sensor to pin B12 of the Blue Pill, as shown in *Figure 9.4*:

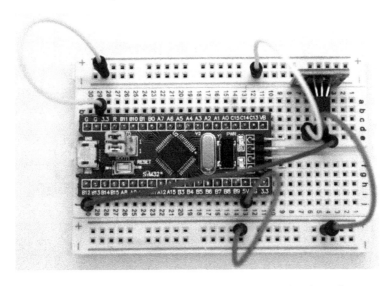

Figure 9.4 – Temperature sensor connection to the Blue Pill

4. Finally, you need to use a power source such as batteries or the STLink connected to the USB port of the computer to power up the board. As usual, we will use the STLink to upload the scripts to the microcontroller board. *Figure 9.5* summarizes all the hardware connections:

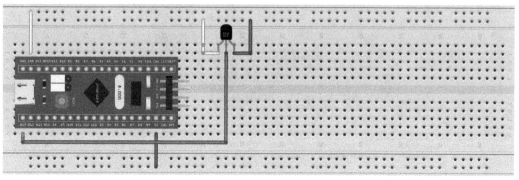

Figure 9.5 – Circuit for the temperature sensor connection

The previous diagram shows all the connections between the STM32 Blue Pill and the electronic components and summarizes the connection steps we just completed.

Figure 9.6 presents the schematics for this project:

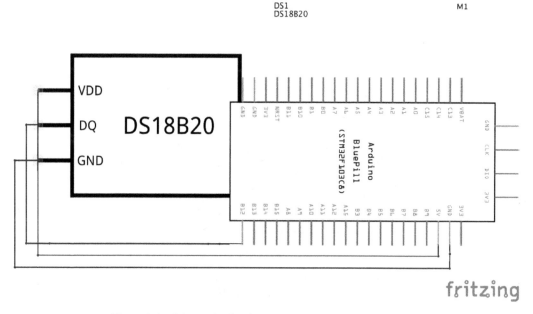

Figure 9.6 – Schematics for the temperature sensor connection

The schematics diagram shows the electric diagram for the complete project. *Figure 9.7* shows how everything is connected in our temperature-logging system:

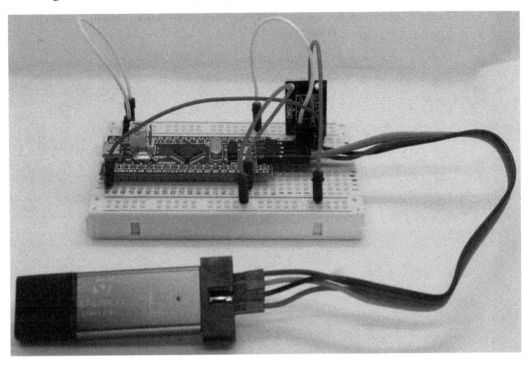

Figure 9.7 – Temperature-logging system

This section introduced you to the DS18B20 temperature sensor. We discovered its specifications and advantages versus other kinds of sensors for measuring temperature. Then you learned how to wire it on the breadboard to the interface with the STM32 Blue Pill.

It is now time to move on to the next section, which will present the C code to complete the first functionality of the IoT temperature logging.

Coding a temperature reading system

In this section, we will develop the program to take temperature readers from a sensor. As mentioned, the DS18B20 sensor works with the 1-wire protocol, so we will use the Arduino IDE libraries to program it. Let's get started:

1. As the first step, we are going to install the **OneWire** library. Open the Arduino IDE, and then go to the **Tools** menu and then **Manage Libraries** (see *Figure 9.8*):

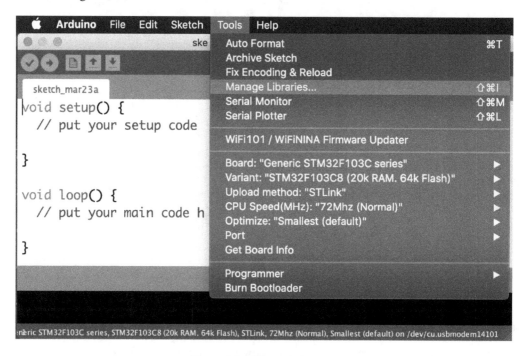

Figure 9.8 – Library manager

2. Next, we will search the library by entering the word OneWire in the search box. We will install the one created by the 1-wire protocol developers, so please install the one from Jim Studt and his colleagues (see *Figure 9.9*):

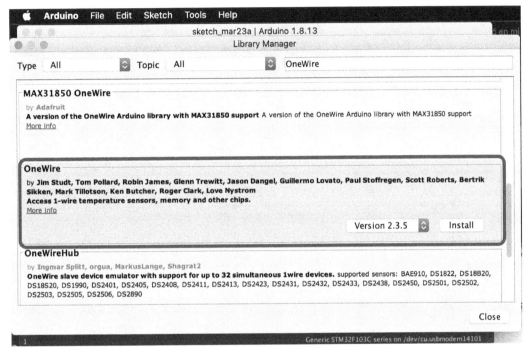

Figure 9.9 – Installing the OneWire library

3. Next, we are going to add the Dallas Temperature library. For this, we enter `ds18b20` in the search box and install the library developed by Miles Burton and collaborators (see *Figure 9.10*). This library is also available from the sensor producers, Dallas Semiconductor (now Maxim):

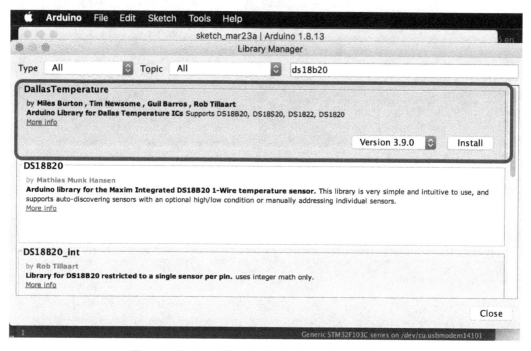

Figure 9.10 – Installing the Dallas Temperature library

Another way to install the libraries without using the built-in function of the Arduino IDE is to download the libraries from their repositories on GitHub manually. After downloading, please place them in the `Arduino/Libraries` folder. Next, find the repositories for the libraries at the following links.

OneWire: `https://github.com/PaulStoffregen/OneWire`.

Maxim (former Dallas) Temperature: `https://github.com/milesburton/Arduino-Temperature-Control-Library`.

4. Let's write the code. We need to include the previously installed libraries and define which pin of the STM32 Blue Pill card pins will be used for input:

```
#include <DallasTemperature.h>
#define PIN_1_WIRE PB12
OneWire pinWire(PIN_1_WIRE);
DallasTemperature sensors(&pinWire);
```

As you can see in the preceding snippet, the One Wire Bus will be PB12 (labeled P12 on the Blue Pill). Also, an instance has been created to carry out the communication, and the instance's reference is passed to the temperature sensor.

5. Next, in the setup() part, we need to start the serial data transmission and assign the speed of the transfer (as usual, we will use 9,600 bps as the standard value):

```
void setup() {
    Serial.begin(9600);
}
```

6. We also need to start reading the sensor:

```
void setup() {
    Serial.begin(9600);
    sensors.begin();
}
```

7. Now comes the loop() part in the sketch. The requestTemperatures() function reads the value of the temperature sensor after reading the value shown on the console:

```
void loop() {
    sensors.requestTemperatures();
    int temp = sensors.getTempCByIndex(0);
    Serial.print("Temperature = ");
    Serial.print(temp);
    Serial.println(" °C");
    delay(1000);
}
```

Important note

If the DS18B20 temperature sensor's read value is -127, this means that something is not well connected. Verify all the wiring. Perhaps the pin connections are wrong or it is simply a case of slack cables.

We have the complete code for reading the temperature. Next, we can see the complete sketch, available in the Chapter9/temperature_reading folder in the GitHub repository.

Now that the sketch is complete, you can upload it to the Blue Pill Board. You can see in the **Serial monitor** the temperature that the sensor is measuring (as shown in *Figure 9.11*):

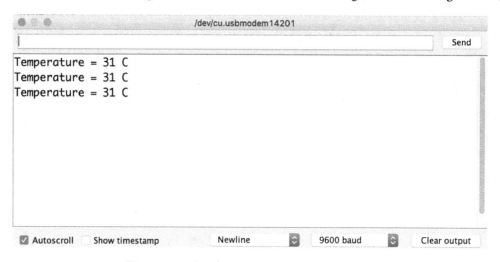

Figure 9.11 – Serial monitor temperature readings

So far, we have learned to measure the environment temperature from a sensor. During the main loop, the device keeps sensing the sensor and displays its data gathered on the serial monitor.

Next, we are ready to learn about the ESP8266 sensor module and how to connect it to load the required scripts for internet connection

Learning to connect the ESP8266 module

As we learned at the beginning of the chapter, an electronic device to be considered an IoT device must have an internet connection and make its data available through this medium.

Due to the aforementioned requirement, we will use a module that will give our temperature-logging system the ability to connect to the internet. This component is the ESP8266 Wi-Fi module.

Now, we are going to learn the hardware components needed to connect the **STM32 Blue Pill** to the internet using the ESP8266 Wi-Fi module. The first thing will be to know and understand the Wi-Fi module.

An introduction to the ESP8266 Wi-Fi module

The **ESP8266** is a microcontroller with integrated Wi-Fi communication, and its main advantage is its very low cost compared to other chips with similar characteristics. By itself, it can work as a microcontroller, such as Arduino or Blue Pill, but it is widely used as a Wi-Fi module for other microcontrollers that do not have a built-in internet connection. This project will use it as the main microcontroller to manage the internet connection and temperature measurement. After the ESP8266 receives a remote interaction from the internet, it will connect to the STM32 Blue Pill to demonstrate the connection between both microcontrollers.

This chapter will use the ESP-01 module, which includes in a breakout board the ESP8266 chip, Wi-Fi antenna, flash memory, LEDs, and pins to connect to solderless breadboards without the need for soldering (as shown in *Figure 9.12*), just with a few jumper wires:

Figure 9.12 – ESP-01 breakout board with an ESP8266 Wi-Fi

The ESP-01 enables Wi-Fi communication using the TCP/IP stack over the serial port using AT commands by default from the factory.

The ESP8266 module has three types of operation:

1. **Station (STA)**.

2. **Access Point (AP)**.

3. Both.

4. In **AP mode**, the module acts as an access point on a Wi-Fi network to connect other IoT devices. In the **STA** type, our modules can be connected to a Wi-Fi access point of a network. The latter mode allows the SP-01 to operate as **AP** and **STA**. In this chapter, we will override the AT firmware of the SP-01 to use firmware coded by ourselves as a C script.

With a knowledge of basic features composing the ESP8266, including the ESP-01 module, let's move on to the following subsection to learn how to connect it to the STM32 Blue Pill.

Connecting an ESP8266 Wi-Fi module

In this subsection, we will learn to interface the ESP-01 for uploading our C scripts. In the end, we will have an electronic device with the ability to connect to a Wi-Fi network.

One of the most critical issues when prototyping with the SP-01 is that the pins are not physically compatible with a solderless breadboard, so we will need jumper cables to accomplish the connections.

The pin configuration of the SP-01 is as follows:

- **GND** corresponds to the ground.
- **GPIO2** general-purpose input-output. It is digital pin number 2.
- **GPIO0** general-purpose input-output. It is digital pin number 0.
- **RXD** is the pin where the serial port data will be received. It works at 3.3 V. It can also be used as a GPIO digital pin. This will be number 3.
- **TXD** is the pin where the serial port data will be transmitted. It works at 3.3 V. It can also be used as a GPIO digital pin. This will be number 1.
- **CH_PD** is the pin to turn the ESP-01 on and off. If we set it to 0 V (LOW), it turns off, and it turns on at 3.3 V (HIGH).
- **RESET** is the pin to reset the ESP-01. If we set it to 0 V (LOW), it resets. VCC is where we feed the ESP-01. It operates at 3.3 V and supports a maximum of 3.6 V. The current supplied must be greater than 200 mA.

As you can see, the SP-01 does not have analog pins, but it does have four digital pins available to work with: GPIO0, GPIO2, RXD (GPIO3), and TXD (GPIO1). To program the SP-01, we require a **USB-Serial adapter** (also called **FTDI** or **TTL**) and, at the same time, we can power it.

Figure 9.13 shows the connections between the SP-01 and the FTDI adapter:

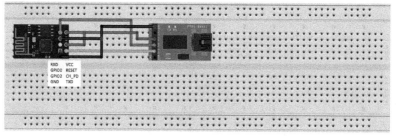

Figure 9.13 – An ESP8266 connected to the Blue Pill

The following are the steps for connecting the ESP8266 to the FTDI, according to the previous diagram:

1. Connect the ESP8266's **RXD** pin to the FTDI's **TXD** pin.

2. Connect the ESP8266's **TXD** pin to the FTDI's **RXD** pin.

3. Connect the ESP8266's **GND** pin to the FTDI's **GND** pin.

4. Connect the ESP8266's **CH_PD** pin to the FTDI's **3.3 V** pin.

5. Connect the ESP8266's **3.3 V** pin to the FTDI's **3.3 V** pin.

To complete *steps 4* and *5*, you will need a solderless breadboard. *Figure 9.14* shows what the connection between the FTDI and ESP-01 module looks like after everything was connected:

Figure 9.14 – Wi-Fi module connections

Important note

To load a program in the ESP-01, we must have the GPIO0 pin at a low level (LOW = GND) and the GPIO2 pin at a high level (HIGH = VCC). We must remember that the ESP8266 works with 3.3 V logic levels, so the GPIO2 pin is HIGH by default since it has an internal pull-up. Therefore, it can be left disconnected.

Finally, *Figure 9.15* shows how all the finished hardware connections appear, including the temperature sensor:

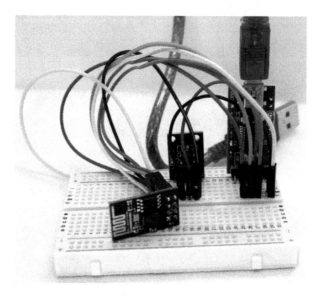

Figure 9.15 – IoT temperature-logging system

Recapping what we have learned in the chapter, we now know how to obtain the ambient temperature using the DS18B20 temperature sensor and the STM32 microcontroller. We met the Wi-Fi module SP-01 and interfaced with an FTDI adapter to program and power it.

It is time to go to the next section, which will present the C code to connect the temperature sensor to the internet using the SP-01.

Coding a program to send the sensed temperature to the internet

Now, we need to develop the software for connecting the temperature sensor to the internet using the ESP8266 Wi-Fi module. Let's begin:

1. Open the **Arduino** menu and select **Preferences**.

2. Add `https://arduino.esp8266.com/stable/package_esp8266com_index.json` to the **Additional Boards Manager URLs** field. You will need to separate the text with a comma from the link of the STM32 module that we installed in the first chapters.

3. Install the **esp8266** platform. Go to the **Tools** menu and select **Board** followed by **Boards Manager** (see *Figure 9.16*):

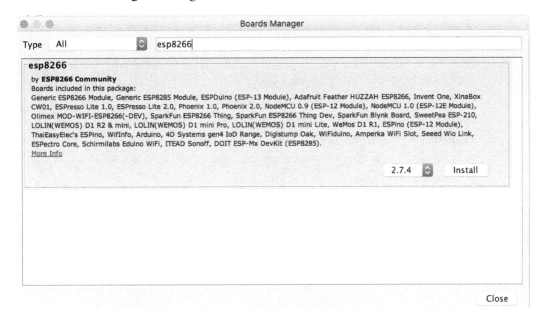

Figure 9.16 – Installing the esp8266 platform

4. Including the libraries will be the first step in the code:

```
#include <DallasTemperature.h>
#include <ESP8266WiFi.h>
#include <ESP8266WebServer.h>
```

5. `setup()` will contain all the programming logic. We need to start the serial data transmission and assign the speed of the transfer (this time we will use 115,200 bps). At the same time, we will initialize the sensor readings:

```
void setup() {
  Serial.begin(115200);
    sensors.begin();
}
```

6. Next, we are going to add the Wi-Fi credentials and start a web server. When the server receives a request to read, this will call a function named `read_sensor`:

```
void setup() {
  Serial.begin(115200);
  sensors.begin();
  WiFi.softAP(ssid, password);
  Serial.print("Connected, IP address: ");
  Serial.println(WiFi.localIP());
  server.on("/", [](){
    Serial.println("Incomming connection to server");
    server.send(200, "text/html", strFrm);
  });
  server.on("/read", read_sensor);
  server.begin();;
}
```

7. When the web server starts, an HTML button will be displayed as a command from the user to read the temperature from the sensor.

```
String strFrm = "<form action='read'><input type='submit'
value='Read sensor'></form>";
```

8. Finally, after the user presses the button, the server will execute the `read_sensor()` function. This function will read the sensor value and display it to the user over the internet:

```
void read_sensor() {
  Serial.print("Reading the sensor: ");
  sensors.requestTemperatures();
  int temp = sensors.getTempCByIndex(0);
  Serial.println(temp);
  server.send(200, "text/plain",
    String("Temperature: ") + String(temp));
}
```

9. The `loop()` part in the sketch will keep the internet connection waiting for the user commands:

```
void loop(void) {
    server.handleClient();
}
```

Now, you can upload it to the ESP8266. To test the program, just open any web browser and open the IP assigned to our device. *Figure 9.17* shows the temperature reading over the internet using the ESP8266 Wi-Fi module:

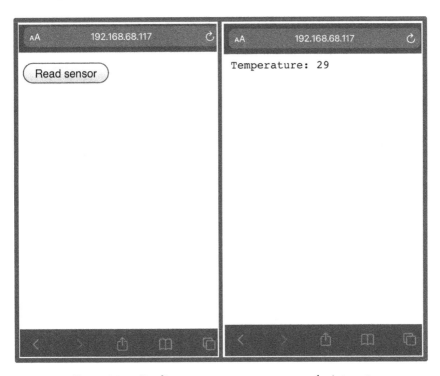

Figure 9.17 – Reading a temperature sensor over the internet

This time, we have the complete code for connecting the temperature sensor to the internet. We can find the complete sketch in the `Chapter9/wifi` folder in the GitHub repository.

We will make the first approach for the STM32 Blue Pill microcontroller to obtain data from the internet in the next section.

Connecting the STM32 Blue Pill board to the internet

The previous code snippets sense a sensor to measure the temperature and send the sensed data to the internet.

When the user requests the temperature from the web browser, the STM32 microcontroller will receive a request to blink an LED and consequently link it to our IoT environment created with the ESP8266 Wi-Fi module.

Figure 9.18 shows the connections required to interface the STM32 and the SP-01:

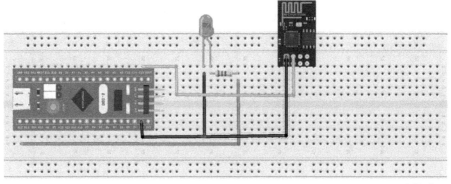

Figure 9.18 – Connecting the STM32 to the internet

Figure 9.19 shows the actual device connections between the STM32 and the SP-01:

Figure 9.19 – Physical connections between the STM32 and SP-01

To complete the connection between the STM32 and the SP-01, we need to add a few lines of code to the Chapter09/wifi script:

```
const int toInternetPin = 0;
```

In the preceding line, add a constant to store the input pin used to receive the data from the internet. Then, in the read_sensor() function, add the following line to send the value 1 (HIGH) each time the user ask for the temperature:

```
digitalWrite(toInternetPin, HIGH);
```

To finish, upload the Chapter09/internetblink script to the STM32 Blue Pill microcontroller in order to read a digital input and send a digital output to blink a LED. This script will not be explained here because it uses a set of instructions that are well-known to the reader.

Open a web browser and go to the IP address of our server and press the **Read sensor** button. You will see the temperature and the LED blinking.

Congratulations! You have finished learning how to connect a temperature sensor to the internet using the ESP8266 Wi-Fi module ESP-01 and how to open a connection between the STM32 Blue Pill and an internet request.

Summary

What have we learned in this project? Firstly, we learned how to connect a temperature sensor to the STM32 Blue Pill microcontroller board. We then wrote the code to read the temperature and send it to our microcontroller. Subsequently, we learned how to connect a Wi-Fi module to our STM32 and code a sketch to connect the board to the internet.

This project has given us the skills to begin to create an IoT application, a great skill in this hyper-connected world. In the forthcoming chapters, you will be able to apply what you have learned since they consist of projects that require an internet connection. In *Chapter 10, IoT Plant Pot Moisture Sensor*, you will learn about measuring the moisture of a pot through a sensor and sending it to the cloud. We will visualize the sensor data on a web page.

Further reading

- Morgan, J. *A Simple Explanation of "The Internet of Things"*. Forbes. 2014: https://www.forbes.com/sites/jacobmorgan/2014/05/13/simple-explanation-internet-things-that-anyone-can-understand/

10
IoT Plant Pot Moisture Sensor

With the advent of the **Internet of Things (IoT)**, we are immersed in the new industrial revolution—the so-called **Industry 4.0**. One of the industries that have benefited the most from these technologies is agriculture (Chalimov, 2020). Agricultural IoT applications range from autonomous harvesting to sensors to recognize pests and diseases or to measure humidity. We can take advantage of these advances in our homes—for instance, monitoring our ornamental plants to have more efficient care.

In this chapter, you will put into practice information learned in *Chapter 9, IoT Temperature-Logging System*, about how to connect and program an internet connection, but this time we will introduce a **NodeMCU microcontroller** to facilitate the ESP8266 programming. You will learn how to create a digital device to monitor a plant pot, reading data from a soil moisture sensor and determining whether it needs water, and then sending an alert to notify that it is too dry.

In this chapter, we will cover the following main topics:

- Connecting a soil moisture sensor to the Blue Pill board
- Reading data from the soil moisture sensor module
- Coding a program to send the sensed data to the internet
- Showing sensor data results over the internet

By completing this chapter, you will discover how to read the soil's moisture amount through a sensor connected to the STM32 Blue Pill board. You will also learn how to send this information to the internet through the NodeMCU development board and visualize the sensor values from a responsive web page.

Technical requirements

The hardware components that will be needed to develop the plant pot moisture system are listed as follows:

- One solderless breadboard.

- One Blue Pill microcontroller board.

- One NodeMCU microcontroller.

- One ST-Link/V2 electronic interface, needed for uploading the compiled code to the Blue Pill board. Bear in mind that the ST-Link/V2 interface requires four female-to-female jumper wires.

- One soil moisture sensor.

- One ESP8266 Wi-Fi module.

- Male-to-male jumper wires.

- Power source.

These components can be easily obtained from your favorite supplier. Additionally, you will require the Arduino **integrated development environment** (**IDE**) and the GitHub repository for this chapter, which can be found at `https://github.com/PacktPublishing/DIY-Microcontroller-Projects-for-Hobbyists/tree/master/Chapter10`

The Code in Action video for this chapter can be found here: `https://bit.ly/3d9CmNM`

The next section presents an introduction to a soil moisture sensor and how to use it with the STM32 Blue Pill microcontroller board.

Connecting a soil moisture sensor to the Blue Pill board

We will start this section by learning how to use a sensor to measure soil humidity in a plant pot, and you will later learn how to connect it to the **STM32 Blue Pill** board to build a plant pot moisture-monitoring system.

Introducing soil moisture sensors

A **soil moisture sensor** consists of two pads that measure the amount of water in the soil. The sensed value is acquired by allowing the the the electric current running through the soil to pass through, and according to resistance, this shows the level of humidity in the plant pot.

You can find a generic breakout module that is pretty straightforward to use. This is shown in the following photo:

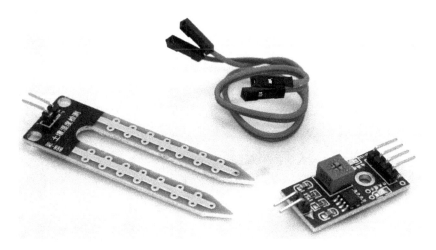

Figure 10.1 – Soil moisture sensor breakout board

The pads are connected to the breakout board with the included female-to-female jumper wires. The breakout board connects to the STM32 Blue Pill board with four pins, outlined as follows:

- **Analog output (AO)**: This pin generates an analog signal and must be connected to an analog input of the microcontroller.

- **Digital output (DO)**: This pin generates a digital signal and must be connected to a digital input of the microcontroller.

- **VCC**: Pin to supply power to the sensor (3.3 **volts** (**V**)-5 V).

- **Ground (GND)**: Ground connection.

To simplify the development of our project, we will use a DO pin to build our system because it only generates binary data depending on the humidity.

Connecting the components

We will use a solderless breadboard to connect the sensor and the STM32 Blue Pill microcontroller, and finally wire to connect the components. Follow these steps:

1. Place the soil moisture sensor and the STM32 Blue Pill board on the solderless breadboard with enough space to add the wiring layer.

2. Connect the ground (GND) pin of the sensor to a GND terminal of the SMT32 Blue Pill board.

3. Next, you need to connect the voltage (VCC) pin to the 3V3 bus of the STM32 Blue Pill board. The sensor DO must be connected to a digital input on the STM32 Blue Pill board, so connect the DO pin of the sensor to pin B12 of the Blue Pill, as shown in the following photo:

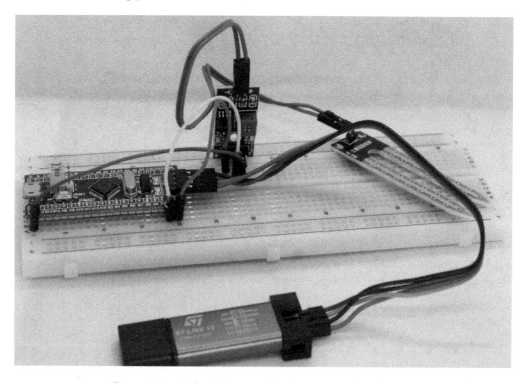

Figure 10.2 – Soil moisture sensor connection to the Blue Pill

4. Finally, you will need a power source to power up the board. Use the ST-LINK to upload the scripts to the STM32 Blue Pill microcontroller board. The following screenshot summarizes all the hardware connections:

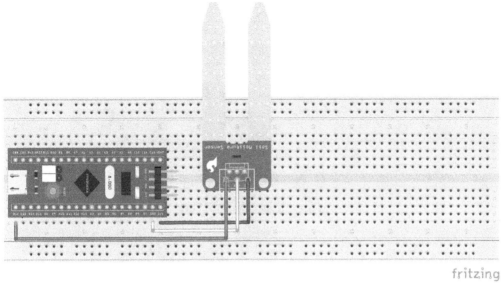

Figure 10.3 – Circuit for soil moisture sensor connection

The following screenshot presents a schematic diagram for this project:

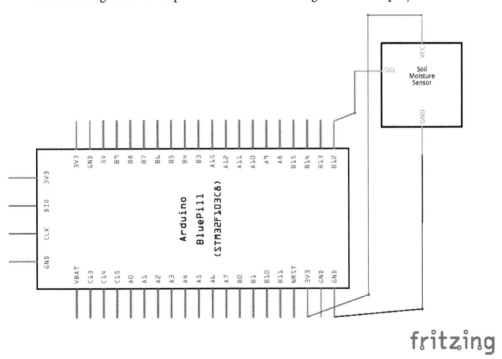

Figure 10.4 – Schematic diagram for soil moisture sensor connection

Figure 10.4 shows an electric diagram for this part of the project. The ground pin of the sensor was connected to the GND pin of the Blue Pill, while the VCC pin was connected to the Blue Pill's 3V3 bus. Finally, the DO of the sensor was plugged into the B12 pin of the STM32 microcontroller. The following photo shows the plant pot moisture system:

Figure 10.5 – Plant pot moisture system

In the previous photo, we can see how the humidity monitoring system's deployment is carried out. As we can see, we built a compact electronic circuit to monitor the moisture of the soil in a plant pot.

In this section, we understood the concept of a humidity sensor and its components. Furthermore, we learned how to connect the sensor to a microcontroller through a breadboard, and finally learned how to connect the complete system to a plant pot.

It's time to move on to the next section, which will show you how to write C code to complete the IoT humidity monitoring system's first functionality.

Reading data from the soil moisture sensor module

You will now learn how to code a program that reads the information from the moisture sensor and shows on the serial monitor if the plant pot needs watering or is moist enough.

Let's start developing the program to collect the sensor data from the STM32 Blue Pill, as follows:

1. Let's get started writing the code. This time, we won't need any additional libraries. Define which of the STM32 Blue Pill card pins will be used as input for reading the sensor data. Also, declare a variable to save the reading data from the sensor, as follows:

    ```
    const int sensorPin = PB12;
    int sensorValue = 0;
    ```

    The input pin will be the PB12 pin (labeled B12 on the Blue Pill). Also, we initialize the sensorValue variable to a value of 0.

2. Next, in the setup() part, we need to start the serial data transmission and assign the speed of the transfer (as usual, we will use 9,600 **bits per second** (**bps**) as the standard value). Here is the code to do this:

    ```
    void setup() {
        Serial.begin(9600);
    }
    ```

3. Indicate to the microcontroller the type of pin assigned to PB12 by running the following code:

    ```
    void setup() {
        Serial.begin(9600);
        pinMode(sensorPin, INPUT);
    }
    ```

4. The rest of the sketch is in the loop() part. The first lines read the input pin's data sensor and display its value in the serial console. The code is shown in the following snippet:

    ```
    void loop() {
        sensorValue = digitalRead(sensorPin);
        Serial.print("Sensor value: ");
    ```

```
    Serial.println(sensorValue);
    if (sensorValue == 1) {
      Serial.println("Soil is too dry");
      delay(1000);
    } else {
      Serial.println("Soil is moist enough");
      delay(1000);
    }
}
```

The value read from the sensor could be 1 or 0; remember, we are reading a digital value. If the value is 1, then the plant pot needs water; otherwise, it is moist enough.

The code is now complete. You can find the complete sketch in the Chapter10/ moisture folder in the GitHub repository.

5. Now that the sketch is complete, you can upload it to the Blue Pill board and insert the sensor pads into a plant pot. Now, you can see in the **serial monitor** that the soil is too dry, as shown in the following screenshot:

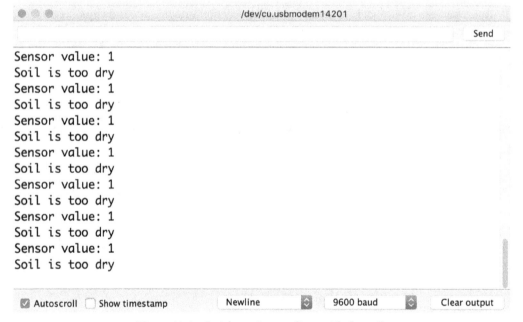

Figure 10.6 – Serial monitor readings with dry soil

6. Now, add water to the plant pot, taking care not to get any electronic components wet. The serial monitor's information will change as soon as the soil gets wet, as illustrated in the following screenshot:

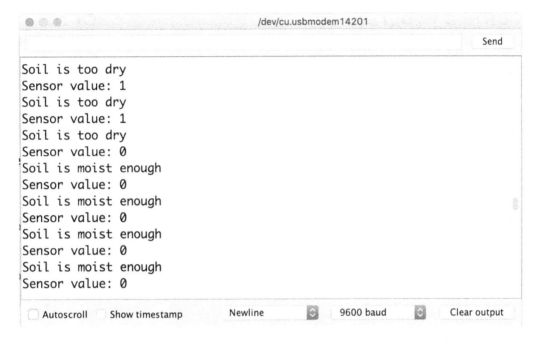

Figure 10.7 – Serial monitor readings upon moistening the soil

> **Important note**
>
> In recent macOS versions, the **Universal Serial Bus** (**USB**) port may not appear in the Arduino IDE, and it therefore may not be possible to see the serial monitor. To solve this, it is necessary to install the USB-UART drivers (where **UART** stands for **Universal Asynchronous Receiver/Transmitter**) from https://www.silabs.com/developers/usb-to-uart-bridge-vcp-drivers.

Let's recap what we have learned so far. We learned about a sensor to measure soil humidity. We learned how to connect it to our STM32 Blue Pill microcontroller in the *Connecting a soil moisture sensor to the Blue Pill board* section. In this section, we wrote the code to obtain its data and display it on the serial monitor.

The skills you have acquired so far in this chapter will allow you to create other electronic systems that require digital reading of data generated in sensors, enabling you to use this sensor in additional projects where it is required to measure soil moisture.

Coming up next, we will learn about the NodeMCU microcontroller, which will facilitate connection to the internet with its integrated ESP8266 module.

Coding a program to send the sensed data to the internet

If you remember, in *Chapter 9, IoT Temperature-Logging System*, we found that an ESP-01 module was used because it integrates Wi-Fi communication through ESP8266. This module was programmed using **AT commands** through the STM32 Blue Pill microcontroller (where **AT** stands for **Attention**). As mentioned at the beginning of the chapter, we will use the **NodeMCU development board**, which is depicted in the following photo:

Figure 10.8 – NodeMCU development board

This board is also based on the ESP8266 microcontroller. However, unlike the SP-01 module, this can be programmed directly from its micro USB port using different development IDEs and various programming languages such as Lua and C. It also includes **general-purpose input/output** (**GPIO**) pins to be programmed according to the developer's needs. These characteristics make the NodeMCU microcontroller one of the most popular IoT platforms today.

The NodeMCU associates with both firmware and development boards, and in conjunction offers the most popular open source IoT platform. The development board is based on the ESP-12 module that, as with the ESP-01 module, gives us the Wi-Fi connection functionality and adds the functionality of the development board, with the following features:

- Micro USB port and serial-USB converter

- Simple programming via micro USB

- Power via USB terminals (pins) for easy connection

- Integrated reset button and **light-emitting diode** (**LED**)

Using its pins, we can easily place it on a solderless breadboard to connect the electronic components required by the projects we will carry out. The NodeMCU enables Wi-Fi communication using the **Transmission Control Protocol/Internet Protocol (TCP/IP)** stack.

Important note

To program the NodeMCU, the steps to add this type of board indicated in the *Showing sensor data results over the internet* section of *Chapter 9, IoT Temperature-Logging System*, must already have been carried out.

Let's create a program to connect the NodeMCU to the internet. Follow these steps:

1. First, include the Wi-Fi library for the ESP8266. You will need two string-type variables for the Wi-Fi network's **service set identifier** (**SSID**) and password (don't forget to change these to your actual values). Also, define an input pin and default value to read the sensor data from the STM32. The possible values gathered from the sensor are listed as follows:

- 0: Moist

- 1: Dry

- 2: Without reading. Hardcoded here, not from the sensor

    The code is illustrated in the following snippet:

```
#include <ESP8266WiFi.h>
const char* ssid = "Your_SSID";
const char* password = "Your_Password";
const int fromStm32Pin = 4;
int sensorValue = 2;
```

2. We will create a web server to receive the sensor data. The server will be listening on port 80. Here is the code to do this:

```
WiFiServer server(80);
```

3. In the `setup()` part, we need to start the serial data transmission and assign the speed of the transfer (this time, we will use 115,200 bps). The code is shown in the following snippet:

```
void setup() {
    Serial.begin(115200);
}
```

4. Indicate to the NodeMCU board the type of pin for reading the STM32, as follows:

```
void setup() {
    Serial.begin(115200);
    pinMode(fromStm32Pin, INPUT);
}
```

5. The rest of the `setup()` part will configure the Wi-Fi network, and upon a successful connection will send the IP address to the serial monitor. The code can be seen here:

```
void setup() {
    Serial.begin(115200);
    pinMode(fromStm32Pin, INPUT);
    Serial.print("Connecting to WiFi network: ");
    Serial.println(ssid);
    WiFi.begin(ssid, password);
    while (WiFi.status() != WL_CONNECTED) {
        delay(500);
        Serial.print(".");
    }
    Serial.println("");
    Serial.println("WiFi connected.");
    Serial.println("IP address: ");
    Serial.println(WiFi.localIP());
    server.begin();}
```

The `loop()` part was built into three functionalities. First, start the web server. Then, read the sensor data from the STM32. Finally, display a responsive web app to visualize the sensor monitoring.

For the complete sketch, refer to the `Chapter10/webserver` folder in the GitHub repository.

6. The web server will be listening for incoming connections from clients. After a client connects, we catch it on the `if` condition, as illustrated in the following code snippet:

```
void loop() {
  WiFiClient client = server.available();
  if (client) {
    // Code to serve the responsive webapp.
  }
}
```

7. After a client connects, the code verifies that is receiving a GET request with a command to read the sensor data, as illustrated in the following code snippet:

```
void loop() {
  WiFiClient client = server.available();
  if (client) {
    if (header.indexOf("GET /4/read") >= 0) {
      Serial.println("Reading the sensor");
      sensorValue = digitalRead(fromStm32Pin);
    }
  }
}
```

8. If the request received by the client asks the sensor value, the NodeMCU will take from the STM32 Blue Pill a reading of the sensor.

To make this bridge between the NodeMCU and the STM32, it will be necessary to add the additional connections shown in the following screenshot:

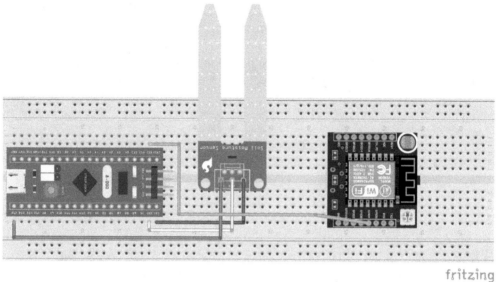

Figure 10.9 – Circuit for microcontrollers' interconnection

Here is a schematic diagram of the microcontrollers' interconnection:

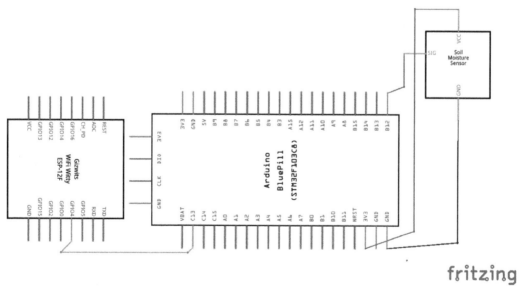

Figure 10.10 – Schematic diagram of microcontrollers' interconnection

Connect a male-to-male jumper wire from NodeMCU GPIO 4 (D2) to the GPIO PC13 pin of the STM32.

The following photo shows how everything was connected in the actual system:

Figure 10.11 – STM32 and NodeMCU connection

9. Now, to complete the connection between the NodeMCU and the STM32, it's necessary to add a few new lines of code to the Chapter10/moisture sketch.

    Add a new constant to store the output pin used to send the data to the NodeMCU, as follows:

    ```
    const int sensorPin = PB12;
    int sensorValue = 0;
    const int toInternetPin = PC13;
    ```

    The output pin will be the PC13 pin (labeled C13 on the Blue Pill).

10. In the setup() part, indicate the pin type for PC13, as follows:

    ```
    void setup() {
      Serial.begin(9600);
      pinMode(sensorPin, INPUT);
      pinMode(toInternetPin, OUTPUT);
    }
    ```

11. Modify the `if` condition in the `loop()` part, as follows:

```
void loop() {
  if (sensorValue == 1) {
    digitalWrite(toInternetPin, HIGH);
    Serial.println("Soil is too dry");
    delay(1000);
  } else {
    digitalWrite(toInternetPin, LOW);
    Serial.println("Soil is moist enough");
    delay(1000);
}}
```

With the preceding code, the STM32 Blue Pill will send the value 1 (HIGH) or 0 (LOW) according to the humidity sensor. Now, we can upload `Chapter10/moisture` to the STM32 and close the sketch and continue working in `Chapter10/webserver`.

12. The final step to complete our web server is to serve a responsive web app after a client request. In this way, any device connected to the same Wi-Fi network and a web browser will be able to access the sensor reading remotely.

But first, we will learn a few concepts of **HyperText Markup Language** (**HTML**): the markup language to create a web page.

A basic structure of an HTML document could look like this:

```
<!DOCTYPE html>
<html>
  <head>
    <meta name="viewport" content="width=device-width,
      initial-scale=1">
    <title>Page Title</title>
  </head>
  <body>
    <h1>A heading</h1>
    <p>A paragraph.</p>
    <img src="anImage.jpg" >
    <button>A button.</button>
```

```
      </body>
    </html>
```

As we build a responsive web app, it is essential to pay attention to the `<meta>` tag with the `name` property that has a `viewport` value. This tag will be responsible for adjusting our app's layout according to the device with which we are browsing, so we can do it from a desktop PC to a mobile device.

To give it the desired visual style, we can do it in two ways: importing a **Cascading Style Sheets** (**CSS**) file or including the styles within `<style></style>` tags, both within the `<head>` tag, as illustrated in the following code snippet:

```
<link rel="stylesheet" href="styleFile.css">
<style>Some styles</style>
```

For our web app, we are going to need a button. If we do not know much about giving CSS visual style, we can use tools freely available on the internet, such as `https://www.bestcssbuttongenerator.com/`, which will visually generate the style CSS of our buttons.

To include HTML code in our sketch, we will use the following sentence:

```
client.println("<html tags>");
```

The code for visualizing the sensor value on our web app prints a paragraph indicating to the user whether the soil is dry or not, and a graphical indicator to better understand our plant pot state, as illustrated in the following code snippet:

```
if (sensorValue == 1) {
  client.println("<p>Soil is too dry</p>");
  client.println("<p><img width=\"50\" height=\"60\"
    src=\"https://raw.githubusercontent.com/
      PacktPublishing/Creative-DIY-Microcontroller-
        Projects/master/Chapter10/images/
          dry_plant.png\"></p>");
} else if (sensorValue == 0)  {
  client.println("<p>Soil is moist enough</p>");
  client.println("<p><img width=\"50\" height=\"60\"
    src=\"https://raw.githubusercontent.com/
      PacktPublishing/Creative-DIY-Microcontroller-
        Projects/master/Chapter10/images/
          green_plant.png\"></p>");
} else {
```

```
client.println("<p>Press the button to read the
    sensor</p>");
}
client.println("<p><a href=\"/4/read\">
    <button class=\"sensorButton\"><i class=\"fas fa-
        satellite-dish\"></i> Read sensor</button>
        </a></p>");
```

To allow the user to gather the sensor reading, we included a button to press each time they need to know their plant's status. Remember—the complete code for this part of the project is available in the `Chapter10/webserver` GitHub folder.

> **Important note**
>
> If you need to use free images and icons, you can find them on the internet repositories such as the following:
>
> `https://pixabay.com/`
>
> `https://fontawesome.com/icons`

The sketch is now complete, so upload it to the NodeMCU board and reset it after completing the upload. Now, you can see in the **serial monitor** the IP address to connect our client, as shown in the following screenshot:

Figure 10.12 – IP address on the serial monitor

It's now time to move on to the next section, which will show you how to visualize the data over the internet.

Showing sensor data results over the internet

Having objects connected to the internet will allow you to access their data from anywhere that has a connection to that network.

This is why we gave our project the ability to become a web server and thus be able to access the state of the plant pot from any web browser.

For this project, access can only be from our Wi-Fi network. To test its operation, we are going to access the developed web app from any mobile or desktop web browser. Proceed as follows:

1. Open a web browser and go to the IP address of our server (see *Figure 10.12*). You should see our landing page to monitor our plant pot, as shown in the following screenshot:

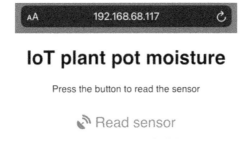

Figure 10.13 – Web app landing page

2. On the landing page, you can simply press the button every time you want to measure the humidity of the plant pot. If the soil is dry, we will see a representative image and a legend stating **Soil is too dry**, as illustrated in the following screenshot:

Figure 10.14 – Web app screen for dry soil

3. Otherwise, if the soil has good humidity, we get a legend stating **Soil is moist enough** along with a representative image, as illustrated in the following screenshot:

Figure 10.15 – Web app screen for moist soil

You can monitor your plant pots with this IoT device and application from anywhere in your home, as long as you are connected to your local Wi-Fi network.

We have reached the end of this chapter. Well done! Let's see what we learned in this project.

Summary

At the beginning of the project, you saw how to interface a soil moisture sensor to your STM32 board. Then, we created a simple sketch to collect the sensor readings and tested it to ensure it worked properly.

We also learned how to connect a NodeMCU card to the internet and read the sensor data from the STM32. Finally, in the last part of the project, we built a web app to control the IoT device from any web browser, either mobile or desktop.

The IoT area is growing quickly, so talented people with the right skills in this technology can easily access jobs in this exciting area. With this in mind, after completing this chapter, we now have a stronger foundation for creating IoT devices and applications.

In the next chapter, we will learn how to connect our electronic devices to the internet and make them available outside our local Wi-Fi network.

Further reading

Chalimov, A, *IoT in agriculture: 8 technology use cases for smart farming (and challenges to consider)*. Eastern Peak, 2020: `https://easternpeak.com/blog/iot-in-agriculture-technology-use-cases-for-smart-farming-and-challenges-to-consider/`

11
IoT Solar Energy (Voltage) Measurement

Solar energy is considered one of the most promising **renewable energy sources** in the face of global warming challenges. It has been considered one of the best alternatives to reduce the dependency on fossil fuels and meet the growing demand for electricity (Ryan, 2005). To achieve this, sunlight is converted into electricity, and the sunlight is collected through solar panels.

In this chapter, you will continue creating IoT software for the STM32 Blue Pill microcontroller board using a voltage sensor to measure the solar energy collected by a solar panel. The application will send the sensed data to the internet using the NodeMCU ESP8266 microcontroller board.

In this chapter, we will cover the following main topics:

- Connecting a solar panel to the Blue Pill board
- Reading data from a voltage sensor module
- Coding a program to send the sensed data to the internet
- Showing sensor data results over the internet

After this chapter, you will have solid skills for developing IoT applications and improving your portfolio because it is a core element in Industry 4.0. The first skill you will learn is reading the solar panel voltage from a sensor connected to the STM32 Blue Pill. Furthermore, you will learn how to send the information read to the internet over the NodeMCU 8266 development board. Finally, you will find out how to visualize sensor values on a mobile IoT application.

Technical requirements

The hardware components that will be needed to develop the solar energy measurement system are as follows:

- One solderless breadboard.
- One Blue Pill microcontroller board.
- One NodeMCU microcontroller.
- One ST-Link/V2 electronic interface for uploading the compiled code to the Blue Pill board. Bear in mind that the ST-Link/V2 requires four female-to-female jumper wires.
- One B25 voltage sensor.
- One solar panel.
- Male-to-male jumper wires.
- Female-to-male jumper wires.
- Power source.

All the components can easily be found at your preferred electronics supplier. Remember, you will require the Arduino IDE and the GitHub repository for this chapter: https://github.com/PacktPublishing/DIY-Microcontroller-Projects-for-Hobbyists/tree/master/Chapter11

The Code in Action video for this chapter can be found here: https://bit.ly/2U4YMsT

The next section presents an introduction to the solar panels and the B25 voltage measurement sensor and how to interface them to the STM32 Blue Pill microcontroller board.

Connecting a solar panel to the Blue Pill board

Firstly, we need to learn about two components: the solar panel and the voltage measurement sensor. After learning the basics, we can build our solar energy measurement system.

Introducing the solar panel

Sunlight carries energy. When sunlight collides with a semiconductor, some energy is changed into moving electrons, generating current. Solar cells (also known as photovoltaic panels or PV panels) were created to take advantage of all the sunlight that reaches our planet. When sunlight reflects off a PV panel, the current output is constant; this is known as **direct current** (**DC**) electricity. This DC can be used to charge batteries and power microcontrollers such as the STM32 Blue Pill.

The following screenshot shows a solar panel for use with electronic components, such as our solar energy demonstration system:

Figure 11.1 – Solar panel

To facilitate the connection and operation with this solar panel, we will solder a pin header to the panel so we can directly connect jumper wires to it. The following figure shows the pin header and how the PV panel looks after being soldered:

Figure 11.2 – Soldering of pin header to the solar panel

You can also find solar panels on the market that already have integrated cables to facilitate their use, such as the one shown in the following figure:

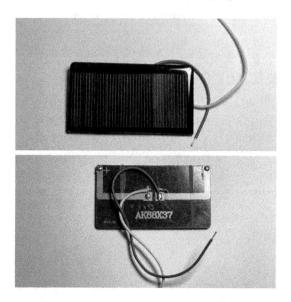

Figure 11.3 – Solar panel with integrated wires

With the knowledge of what a solar panel looks like and its functionality, let's move on to the following subsection, where we will explore the sensor we will use to measure voltage.

The B25 voltage sensor

If we need to measure voltage, we can use the analog inputs of our STM32 Blue Pill board. These inputs have a limit of 5V, so if it is necessary to measure higher voltages, it is necessary to use an external sensor to do it. The **B25 sensor** (see *Figure 11.4*) measures voltages in the 5V to 25V range, making it a very popular sensor for this task:

Figure 11.4 – B25 voltage sensor breakout board

As can be seen, the module has two terminals to which the external power source will be connected, one to GND and the other to VCC, which must be adjusted with a screw.

Additionally, the breakout board connects to the STM32 Blue Pill with 3 header pins. They are as follows:

- **S**: This pin generates an analog signal and must be connected to an analog input of the microcontroller.

- +: Not connected.

- -: Ground connection.

With this information in mind, we will learn how to connect the voltage sensor to the STM32 Blue Pill board in the next subsection.

Connecting the components

We will use a solderless breadboard to connect the sensor and the STM32 Blue Pill microcontroller and finally wire to connect the components. Here's how we wire and connect the components:

1. Place the voltage sensor and the STM32 Blue Pill on the solderless breadboard. Leave some empty space to add the wires.

2. Connect the **ground (GND)** pin of the sensor to a GND terminal of the STM32 Blue Pill.

3. Next, you need to connect the sensor analog output to an analog input on the STM32 Blue Pill card and connect the S of the sensor to pin **A0** of the Blue Pill, as shown in *Figure 11.5*:

Figure 11.5 – Voltage sensor connection to the Blue Pill

4. Finally, you will need a power connection to connect the solar panel to the board. Use the STLink to upload the scripts to the STM32 Blue Pill microcontroller board. *Figure 11.6* summarizes all the hardware connections:

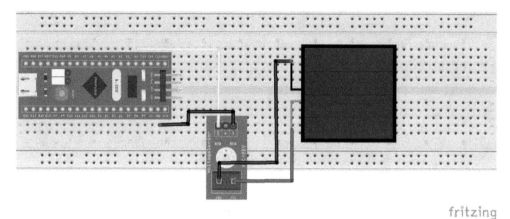

Figure 11.6: Circuit for voltage sensor connection

The following screenshot presents the schematics for this project:

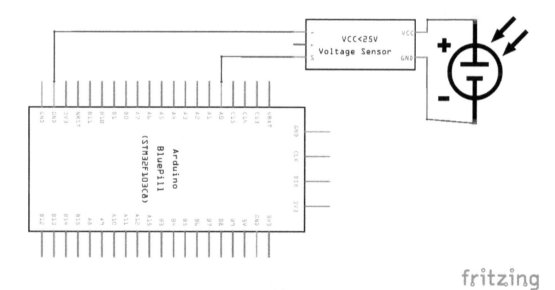

Figure 11.7 – Schematics for voltage sensor connection

The schematics diagram shows the electrical connections. The VCC and ground terminals of the PV panel were connected to the VCC and GND pins of the sensor. To interface the Blue Pill with the sensor, its ground pin was connected to the GND bus of the Blue Pill, and finally, the analog output (**S**) of the sensor was plugged into the pin A0 of the STM32 microcontroller. *Figure 11.8* shows the solar energy measurement system:

Figure 11.8 – Solar energy measurement system

Now that we have finished connecting the components, we have created a simple circuit for our voltage measurement system, as observed in the previous figure.

In this section, we learned about solar panels and met a voltage sensor and its components. We also learned how to connect the solar cell to the voltage sensor and the voltage sensor to the STM32 Blue Pill.

It is time to move on to the next section, which will show you how to write C code to complete our IoT solar energy monitoring system's first functionality.

Reading data from a voltage sensor module

It is time to learn how to code a program that will read the information from the voltage sensor and display its reading on the serial monitor.

Let's write the program to receive the sensor data from the STM32 Blue Pill:

1. Declare which pin of the STM32 Blue Pill card will be used as input of the sensor data:

```
const int sensorPin = 0;
```

The input pin will be the 0 (labeled A0 on the Blue Pill).

2. Next, in the setup() part, start the serial data transmission and assign the speed of the transfer to 9600 bps, and indicate to the microcontroller the type of pin assigned to A0:

```
void setup() {
    Serial.begin(9600);
    pinMode(sensorPin, INPUT);
}
```

3. Now, in loop(), first read the input pin's data sensor, send its value to the serial port, and wait for a second:

```
void loop() {
    int sensorValue = analogRead(sensorPin);
    Serial.print("Voltage: ");
    Serial.println(sensorValue);
    delay(1000);
}
```

4. We are going to load the program to the STM32 board and review the serial plotter of the Arduino IDE to know the waveform of the analog signal that we are reading from the sensor, and the result can be seen in *Figure 11.9*:

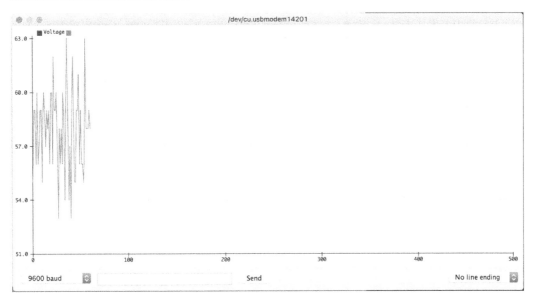

Figure 11.9 – Sensor signal waveform in the serial plotter

The waveform that forms the sensor signal can take the values from 0 to 1023. Then, it will be necessary to convert this value to voltage.

5. We will add two lines to our script to show the voltage value, and we will comment on the one that sends the signal value to the serial port:

```
void loop() {
    int sensorValue = analogRead(sensorPin);
    double voltageValue = map(sensorValue, 0, 1023, 0, 25);
    Serial.print("Voltage: ");
    //Serial.println(sensorValue);
    Serial.println(voltageValue);
    delay(1000);
}
```

The map() function transforms a number from one range to another:

```
map(value, fromLow, fromHigh, toLow, toHigh)
```

The first parameter that map() receives is the value to be converted. In our program, it is the value read from the sensor. The value of fromLow will be mapped to toLow, and fromHigh to toHigh, and all values within the range.

Now, upload it to the Blue Pill board. Now you can see in the **serial monitor** the voltage value as shown in *Figure 11.10*:

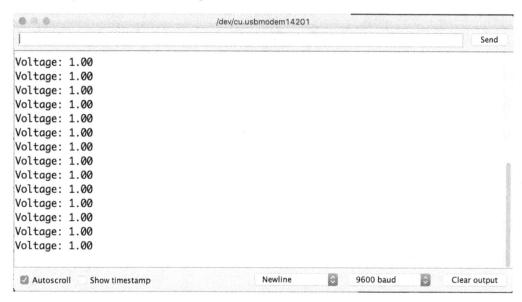

Figure 11.10 – Serial monitor readings

For the complete sketch, refer to the `Chapter11/voltage` folder in the GitHub repository.

What have we learned so far? We introduced the B25 sensor to measure voltage and know about solar panels. We learned to connect them to our STM32 Blue Pill microcontroller, write the code to read the sensor data, display it on the serial monitor, and graph it in the serial plotter.

Some new skills were acquired in this section, and these skills will help you build electronic systems that require monitoring of the voltage level.

Next, we will use the NodeMCU microcontroller to send the sensed data to the internet.

Coding a program to send the sensed data to the internet

In this section, we will continue using the NodeMCU development board to receive the data from the STM32 and send it to the internet. However, unlike *Chapter 10, IoT Plant Pot Moisture Sensor*, where a digital value (1 or 0) was sent directly between both microcontrollers, we now need to send the voltage value using serial communication between these microcontrollers.

Serial transmission is done by sending the data using the RX/TX pins.

Let's create the program to connect the NodeMCU and the STM32:

1. In setup(), we need to add new serial data transmission to 115200 bps. It is the recommended speed for the NodeMCU board:

```
void setup() {
  serial.begin(9600);
  Serial1.begin(115200);
}
```

2. The loop() instance needs a new line after the sensor reading and voltage conversion. The write() function sends the data as an integer value:

```
void loop() {
  int sensorValue = analogRead(sensorPin);
  double voltageValue = map(sensorValue, 0, 1023, 0, 25);
  Serial.print("Voltage: ");
  //Serial.println(sensorValue);
  Serial.println(voltageValue);
  Serial1.write((int)voltageValue);
  delay(1000);
}
```

3. To complete the communication between the NodeMCU and the STM32, it will be necessary to add the additional connections shown in *Figure 11.11* and *Figure 11.12*:

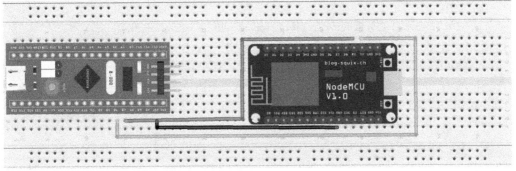

Figure 11.11 – Circuit for microcontroller serial communication

Figure 11.12 shows the schematics diagram for the circuit interfacing between the STM32 and the NodeMCU microcontrollers:

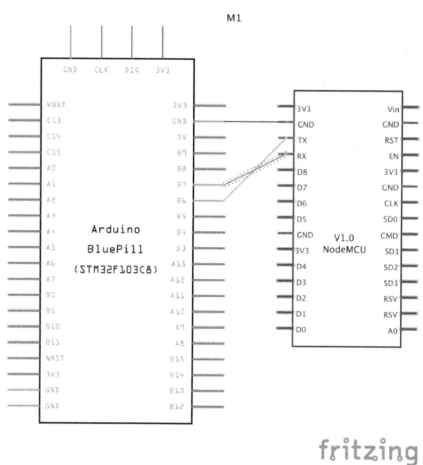

Figure 11.12 – Schematics for microcontroller serial communication

Connect the RX pin from NodeMCU to the TX pin (B6) of the STM32 and the TX pin from NodeMCU to the RX pin (B7) of the STM32.

Figure 11.13 shows how everything was connected in the actual system, including the voltage sensor:

Figure 11.13 – STM32 and NodeMCU serial connection

4. Now, to complete the serial connection between the NodeMCU and the STM32, we will create a new sketch, `Chapter11/voltage_iot`.

5. In `setup()`, indicate the serial data transmission:

```
void setup() {
    Serial.begin(115200);
}
```

6. The final step is `loop()`:

```
void loop() {
    double data = Serial.read();
    Serial.print("Voltage: ");
    Serial.println(data);
    delay(1000);
}
```

With the preceding code, the NodeMCU will receive the sensor value from the STM32 and will display it on the serial monitor.

The sketch is now complete. Upload it to the NodeMCU board, and reset it after completing the upload. Now you can see, in the **serial monitor**, the sensor value, as shown in the following screenshot:

It is now time to move on to the next section, which will show you how to visualize data over the internet.

Showing sensor data results over the internet

In *Chapter 9*, *IoT Temperature-Logging System*, and *Chapter 10*, *IoT Plant Pot Moisture Sensor*, we learned how to program IoT applications within our local network. In this section of the chapter, we will learn how to send data to the cloud outside of our local network.

A wide variety of cloud platforms allow us to connect our IoT devices to their services. Most allow us to use essential services at no cost. If something more complete is desired, there is a charge, generally a monthly payment. This time we will use the Blynk platform, which has several free options, and they are the ones we will use.

Blynk has an app for both Android and iOS that will allow us to monitor the value of the voltage in our solar cell.

Let's look at the steps to send and view our information from the internet with a mobile app:

1. Download the Blynk app.

    For Android, download it from `https://play.google.com/store/apps/details?id=cc.blynk&hl=en_US`.

    For iOS, download it from `https://apps.apple.com/us/app/blynk-iot-for-arduino-esp32/id808760481`.

2. Create a new account:

Figure 11.14 – Blynk, home screen

3. Once your account is created, create a new project. Write a name, select ESP8266 as the device, and set WiFi as the connection type. Then click on **Create Project**:

Figure 11.15 – Blynk, creating a new account

4. You will receive an email with the necessary token for the app, which you can also find in **Settings**:

Figure 11.16 – Blynk, menu screen

5. Write a name, select **ESP8266** as the device, and **WiFi** as connection type. Click on **Create Project**:

Figure 11.17 – Blynk, creating a new project

6. You will receive an email with the necessary token for the app, which you can also find in **Settings**.

7. Press the screen and the Widget toolbox will appear:

Figure 11.18 – Blynk, widgets box

8. Add a **Gauge** component. Configure it and press the **OK** button:

Figure 11.19 – Solar energy app in Blynk

9. Finally, upload the `Chapter11/voltage_iot` program to the NodeMCU and execute it.

We have reached the end of *Chapter 11, IoT Solar Energy (Voltage) Measurement.* Congratulations!

Summary

In this chapter dedicated to IoT, we have learned some essential topics. First, we got to know the solar cells used to power small electronic devices. Next, we learned about the B25 voltage sensor and how to connect it to the STM32.

Later, we learned how to create a program to read data from the voltage sensor. With the voltage reading, we connect our STM32 to a NodeMCU board through serial communication. We create a program to send the voltage value between microcontrollers. Finally, we use an app to visualize the sensor data from the cloud.

At the end of the IoT topics, you have solid skills to create applications and devices connected to the internet and intranets. Your portfolio of projects has been strengthened to enable you to more easily find a job opportunity in this growth area.

In the next chapter, you will start developing projects that will help you create electronic support devices to assist with the COVID-19 pandemic.

Further reading

Ryan, V., *What Is Solar Energy?* Technology Student, 2005:
`https://technologystudent.com/energy1/solar1.htm`

12
COVID-19 Digital Body Temperature Measurement (Thermometer)

This chapter describes an interesting project where you will develop a touchless thermometer to measure human body temperature. This digital thermometer could be useful for supporting the diagnosis of people with COVID-19. The electronics project explained in this chapter involves the use of a very capable **infrared** (**IR**) temperature sensor that will check the body temperature. Additionally, you will learn and practice how to connect an IR temperature sensor to a microcontroller board using the **Inter-Integrated Circuit** (**I2C**) data transmission protocol.

> **Important note**
>
> The body temperature measurement project described in this chapter should not be used as a definitive and accurate way to determine whether a person has COVID-19 or not. It is for demonstration and learning purposes only.

In this chapter, we will cover the following main topics:

- Programming the I2C interface
- Connecting an IR temperature sensor to the microcontroller board
- Showing the temperature on an LCD
- Testing the thermometer

By the end of this chapter, you will have learned how to get useful data from an IR temperature sensor and how to effectively show body temperature data on an LCD connected to a microcontroller board. You will also learn how the I2C data transmission protocol works to get data from the IR sensor, and how to properly test an IR temperature sensor.

Technical requirements

The software tool that you will be using in this chapter is the **Arduino IDE** for editing and uploading your programs to the Blue Pill microcontroller board. The code used in this chapter can be found in the book's GitHub repository:

```
https://github.com/PacktPublishing/DIY-Microcontroller-
Projects-for-Hobbyists/tree/master/Chapter12
```

The Code in Action video for this chapter can be found here: `https://bit.ly/2SMUkPw`

In this chapter, we will use the following pieces of hardware:

- One solderless breadboard.
- One Blue Pill microcontroller board.
- One micro-USB cable for connecting your microcontroller board to a computer and a power bank.
- One Arduino Uno microcontroller board.
- One USB 2.0 A to B cable for the Arduino Uno board.
- Two USB power banks.
- One ST-Link/V2 electronic interface, needed for uploading the compiled code to the Blue Pill. Bear in mind that the ST-Link/V2 requires four female-to-female DuPont wires.
- One MLX90614ESF-DCA-000 temperature sensor (it works with 3.3 volts).

- One 0.1 microfarad capacitor. It generally has a 104 label on it.

- One 1602 16x2 I2C LCD.

- A dozen male-to-male and a dozen male-to-female DuPont wires.

The next section describes how to code the I2C protocol and the code that will run on the Blue Pill and Arduino Uno microcontroller boards. The Arduino Uno is used for getting data from the IR sensor.

Programming the I2C interface

In this section, we will review how to obtain useful data from the MLX90614 temperature sensor to be transmitted using the **I2C** protocol, also known as **IIC**. It is a serial data communication protocol that is practical for interfacing sensors, LCDs, and other devices to microcontroller boards that support I2C. The next section defines what I2C is.

The I2C protocol

I2C is a synchronous serial communication protocol that allows interconnecting sensors, microcontrollers, displays, **analog-to-digital converters** (**ADCs**), and so on, at a short distance using a common **bus** (a bus works as a main digital road). The I2C bus is composed of a few lines (wires) that all the devices share and use for transmitting and exchanging data. The I2C protocol is practical and beneficial because it uses only two wires for data communication. Another benefit of I2C is that in theory, it can support up to 1,008 devices connected to the same I2C bus! It is also worth mentioning that more than one microcontroller can be connected to the same bus, although they must take turns to access data from the I2C bus. *Figure 12.1* shows an overview of the I2C bus configuration:

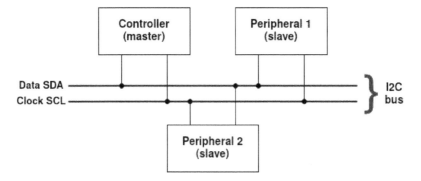

Figure 12.1 – An overview of the I2C bus

As *Figure 12.1* shows, the bus allows the connection of two main types of devices: a **controller** (also known as a **master**) and a **peripheral** (also known as a **slave**). The controller is usually a microcontroller board, but it can also be a personal computer or another type of device that will take, send, process, and use data from and to the peripheral(s). A peripheral can be a sensor that will provide data to a controller, or a display (such as an LCD) where data coming from a controller is displayed. There are other types of peripherals that can be connected to an I2C bus.

> **Note**
> Not all LCDs can be directly connected to an I2C bus. To do that, the LCD must have an I2C adapter connected to it, often called an **I2C backpack** or **I2C module**. This is a small electronic circuit that handles I2C communications, typically attached at the back of some LCDs.

An I2C bus contains two data lines (wires) called **SDA** and **SCL**. The SCL wire transports the clock signal necessary for synchronizing the transmitted data on the bus, and the SDA wire is the data signal transporting all the data between the controller(s) and the peripheral(s).

The devices connected to the I2C bus also have two more wires. One of them is the ground (sometimes labeled GND or Vss). This wire should be connected to the common ground of the electronic circuit where the microcontroller board is connected to, and a voltage wire (labeled Vdd). This is connected to 5 volts, but it sometimes works with 3.3 volts. The I2C protocol is quite robust, allowing a bit rate of up to 5 Mbit/s. Generally, either 5 or 3.3 volts are provided by the microcontroller board, used by the devices connected to the I2C bus.

Many microcontroller boards support the I2C protocol. Fortunately, the Blue Pill includes I2C pins for that. In fact, the Blue Pill has three sets of pins for connecting three I2C devices directly. In this chapter, we will use the Blue Pill's B6 (SCL) and B7 (SDA) pins for the I2C communication only.

> **Note**
> The Arduino microcontroller boards also support the I2C protocol. For example, the pins A4 and A5 from the Arduino Uno board provide the connection for the SDA and SCL wires, respectively.

It is worth noting that the I2C protocol is handled by the `Wire.h` library, found by default in the Arduino IDE configuration. You don't need to install this library.

In the next section, we will review how to code the I2C protocol to get temperature data from the IR sensor using an *Arduino Uno* microcontroller board.

I2C coding

In this section, we review the code for reading the temperature data from the MLX90614 IR sensor by an Arduino Uno microcontroller board, working as a peripheral (slave). This board will send the temperature data to the Blue Pill through the I2C bus, and the Blue Pill (working as a controller) will display it on an LCD. This LCD is also connected to the I2C bus. The full description of the Blue Pill and the Arduino Uno connections is found in the *Connecting an IR temperature sensor to the microcontroller board* section.

The next section explains the necessary code that runs on the Arduino Uno board.

Coding the Arduino Uno software (peripheral)

In order to code an application on the Arduino Uno for reading the MLX90614 IR sensor data, we will use a library to be included in our Arduino IDE program, called `Adafruit_MLX90614.h`. You can install this library from the Arduino IDE:

1. Go to **Tools | Manage Libraries**.

2. Set **Type** as **All** and **Topic** as **All**.

3. In the search field, type `Adafruit MLX90614`.

Install the latest version of the `Adafruit MLX90614` library (do not install the Mini one). The `Wire.h` library controls the I2C protocol, which is already installed on the Arduino IDE's files.

The following is the code that runs on the Arduino Uno (its file is called `peripheral. ino`; you can find it on the GitHub page). `0X8` is the hexadecimal address assigned to the Arduino Uno board as the peripheral (slave) for the I2C protocol. We assigned 0x8 arbitrarily; it can be any hexadecimal address, but make sure to use the same address for both the master and the slave:

```
#include <Wire.h>
#include <Adafruit_MLX90614.h>
Adafruit_MLX90614 mlx = Adafruit_MLX90614();
#define SLAVEADDRESS 0x8
float AmbientobjC=0.0;
```

This function sets up the Arduino Uno as a peripheral, assigning it the 0x8 address. The Blue Pill will identify the Arduino Uno with this address. This function also sets up the interrupt that handles incoming requests from the controller (master), which is the Blue Pill. The Blue Pill will be acquiring the temperature data from the sensor:

```
void setup() {
   Wire.begin(SLAVEADDRESS);
   Wire.onRequest(requestEvent);
}
```

This function will read the temperature from the sensor continuously. The delay function makes a small pause to allow the sensor to get new temperature readings:

```
void loop() {
   AmbientobjC=mlx.readObjectTempC();
   delay(100);
}
```

This function runs every time the controller (master) requests data from the peripheral (slave):

```
void requestEvent() {
   union floatToBytes {
      char buffer[4];
      float objtempReading;
   } converter;
   converter.objtempReading = AmbientobjC+3;
   Wire.write(converter.buffer, 4);
}
```

From the previous code snippet, we see that the write() function is used to send the temperature data to the controller (the Blue Pill).

We added the value of 3 to the AmbientobjC variable to compensate for the ambient temperature. It is important to clarify that the temperature readings are not absolute and will slightly change depending on a number of factors, including the ambient temperature, whether the person is outside. So, you may need to test the temperature readings a number of times and adjust the AmbientobjC variable accordingly, perhaps comparing the sensor readings against medical body thermometer readings.

> **Note**
>
> The code uploaded to the GitHub platform contains extensive comments that explain most of the code lines.

The next section explains the necessary code for running the Blue Pill as a controller.

Coding the Blue Pill software (controller)

The following code (its file is called `controller.ino`; you can find it on the GitHub page) will run on the Blue Pill microcontroller board (the controller). This code will serve to obtain the temperature data sent by the the Arduino Uno board, and then display it on an LCD connected to the I2C bus:

```
#include <LiquidCrystal_I2C.h>
LiquidCrystal_I2C lcd(0x27, 16, 2);
#include <Wire.h>
#define SLAVEADDRESS 0x8
```

The previous code snippet shows the `LiquidCrystal_I2C.h` library for controlling the LCD using the I2C protocol. Its next line sets up the LCD address to `0x27` for a 16-character and 2-line (16x2) LCD. The library can be downloaded from `https://github.com/fdebrabander/Arduino-LiquidCrystal-I2C-library`. Download the `LiquidCrystal_I2C.h` file and copy it to the Arduino libraries folder, usually `Arduino/libraries`.

The next code snippet starts an I2C connection using a peripheral (slave) address, initializing the Serial Monitor and the LCD:

```
void setup() {
   Wire.begin(SLAVEADDRESS);
   Serial.begin(9600);
   lcd.begin();
   lcd.backlight();
}
```

This `loop()` function continuously reads the temperature data sent by the peripheral (the Arduino Uno):

```
void loop() {
   Wire.requestFrom(8, 4);
   uint8_t index = 0;
```

```
union floatToBytes {
    char buffer[4];
    float objtempReading;
} converter;
while (Wire.available()){
    converter.buffer[index] = Wire.read();
    index++;
}
Serial.println(converter.objtempReading);
lcd.setCursor(0, 0);
lcd.print("Body Temp.:");
lcd.print(converter.objtempReading);
delay(500);
}
```

The previous code snippet shows how to read the data in bytes from the Arduino Uno. Remember that we can't transmit float values directly over the I2C bus. The code that runs on the Arduino Uno converts each floating-point temperature data reading into four character bytes. The code running on the Blue Pill converts the four bytes back to a floating-point number.

The code submitted to the GitHub page contains many comments explaining the code lines.

The following section explains how to connect the IR sensor to the Blue Pill microcontroller board.

Connecting an IR temperature sensor to the microcontroller board

This section explains the main technical characteristics of the MLX90614 temperature sensor. It also shows how to connect it to the Blue Pill microcontroller board using the *I2C protocol*.

First, let's explain the main characteristics of the MLX90614 sensor.

The MLX90614 IR sensor

The MLX90614 sensor, manufactured by the company Melexis, is a powerful yet compact IR sensor. This sensor uses IR rays to measure the amount of heat generated by the human body or by basically any object. Being a non-contact thermometer, it reduces chance of spreading disease when checking the body temperature, and you don't need to clean it.

The MLX0-614 is technically a sensor contained in an **integrated circuit** (**IC**), since it has extra electronic components and smaller circuits, including an ADC, a voltage regulator, and a **digital-signal processor** (**DSP**).

The following are some of the sensor's technical characteristics:

- Body temperature range of -40 to +125 degrees Celsius.

- Ambient temperature range of -70 to 382.2 degrees Celsius.

- Medical (high) accuracy calibration.

- Measurement resolution close to 0.02 degrees Celsius.

- This sensor is available in both 3-volt and 5-volt versions.

- A convenient sleep mode that reduces power consumption.

The MLX90614 family of sensors' datasheet can be downloaded from the following link:

```
https://www.melexis.com/-/media/files/documents/datasheets/
mlx90614-datasheet-melexis.pdf
```

The MLX90614 IR sensor generates two types of temperature measurements: ambient temperature and an object temperature reading. The **ambient temperature** is the temperature registered on the IR-sensitive part of the sensor (an internal component), which is close to room temperature. The **object temperature** measures how much IR light is emitted by an object, which can be used for measuring body temperature. In this chapter, we will use only the object temperature measurement.

Figure 12.2 shows the MLX90614 sensor:

Figure 12.2 – The MLX90614 sensor showing its four pins

In *Figure 12.2*, please note that there is a round transparent window on the sensor, where the IR light passes through, hitting an internal sensitive part. This part will convert IR light into electrical impulses.

Figure 12.3 shows the *top* view of the MLX90614 sensor diagram pinout:

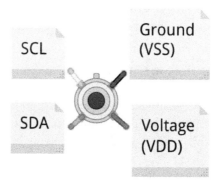

Figure 12.3 – MLX90614 IR sensor pinout

In *Figure 12.3*, you will notice that there is a small notch on top of the sensor. This notch helps you identify which pin is which.

The MLX90614's SCL and SDA pins are connected to a microcontroller board's SCL and SDA, respectively. The ground pin is connected to the microcontroller board's ground, and the voltage pin is connected to either 3.3 or 5 volts.

The MLX90614 sensor is manufactured in different types or versions. One of the most important differences is its supply voltage. It is important to verify its part number:

- **MLX90614ESF-Axx, MLX90614KSF-Axx**: Their supply voltage is 5.5 volts.

- **MLX90614ESF-Bxx, MLX90614ESF-Dxx**: Their supply voltage is 3.6 volts.

For example, the IR sensor that we use in this chapter is the MLX90614ESF-DCA-000, which, according to the manufacturer's datasheet, requires 3.6 volts to work. So, you can use the 3.3 volts supplied by many microcontroller boards for using this type of sensor.

> **Important note**
> Always check the IR sensor part number to determine what type of supply voltage it will need. If you apply a voltage higher than its required supply voltage, you can damage the sensor.

Another technical aspect that you should take into account is the sensor's field of view. It is a relationship between the distance between the sensor and an object that its temperature is being measured. This will determine the sensing area being observed by the sensor. For every 1 centimeter that the object moves away from the sensor's surface, the sensing area grows by 2 centimeters. Ideally, the distance between the sensor and the object (for example, human skin) to be measured should be between 2 and 5 centimeters.

The next section explains the GY-906 module, containing an IR sensor.

The GY-906 sensor module

This section describes the main characteristics and pinout of the GY-906 sensor module. The MLX90614 sensor (see *Figure 12.4*) is also sold embedded in a module called **GY-906**:

Figure 12.4 – The GY-906 module

As *Figure 12.4* shows, the GY-906 module has four pins dedicated to I2C communication. The VIN pin is connected to voltage, either 3.3 volts or 5 volts depending on the type of MLX90641 sensor that it contains. Please refer to the *The MLX90614 IR sensor* section to help identify what voltage the embedded sensor will need in the module. You will also need to consult the module's datasheet. For example, a GY-906 module may have an MLX90614ESF-DCA-000 IR sensor embedded in it, needing 3.3 volts to work. So, GY-906 modules could have any MLX90614 sensor type embedded in them. The module's GND pin is connected to the microcontroller board's ground. The SCL pin is for transmitting the signal clock, and the SDA pin transmits the signal data.

Besides the MLX90614 sensor, the GY-906 module also contains other electronic components, such as pull-up resistors.

The decision on using either an MLX90614 sensor or a GY-906 module rests on a number of factors, including cost, type of application, and size. Bear in mind that the GY-906 module is somewhat larger than the MLX90614 sensor. In this chapter, we will use the MLX90614 sensor and not the module, because the sensor alone is a cost-effective option and to demonstrate how to directly connect the sensor to an I2C bus. However, both the GY-906 module and the MLX90614 sensor have the same functions.

The next section shows how to connect the MLX90614 sensor to a microcontroller board.

Connection of the IR sensor to an Arduino Uno

In this section, we describe how to connect the MLX90614 sensor to the I2C pins of an Arduino Uno microcontroller board, which will be working as a peripheral (slave). The Blue Pill board will be the controller (master), receiving temperature data sent by the Arduino Uno through the I2C bus, as shown in *Figure 12.5*:

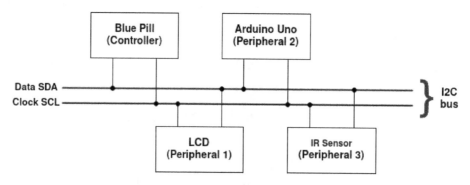

Figure 12.5 – The controller and peripherals

As *Figure 12.5* shows, both the Blue Pill and the Arduino Uno boards are connected to the I2C bus.

We connected the MLX90614 to an Arduino Uno board working as a peripheral (slave) for the following reasons:

- For taking advantage of the I2C data transmission protocol, which only requires two data wires.

- For practicing the implementation of the I2C protocol using controller (master) and peripheral (slave) devices.

- For using MLX90614 software libraries that are fully compatible with other microcontroller boards (for example, the Arduino family), but not with the Blue Pill board. That is why we are connecting the IR sensor to an Arduino Uno board using its respective library. This way we could use the Arduino Uno as a peripheral (slave) that is 100% compatible with the IR sensor.

- For freeing up the Blue Pill (the controller) from processing the sensor's data directly, so the Blue Pill can be used to do other processing-intensive tasks. In addition, the Blue Pill can be dedicated to obtain the sensor's data and display it on an LCD.

First, we will explain how to connect the MLX90614 IR sensor to an Arduino Uno microcontroller board. The connections are shown in *Figure 12.6*:

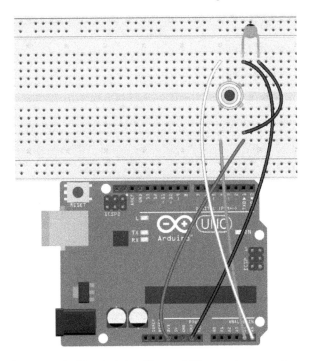

Figure 12.6 – The Arduino Uno and IR sensor's connections

As you can see from *Figure 12.6*, the sensor is connected to analog ports A4 and A5, which are also the Arduino Uno's I2C pins. The 0.1 microfarad capacitor is recommended by the sensor's datasheet to smooth out any high-frequency or very-high frequency electrical noise that may be present in the sensor that may affect the temperature readings. Here are the steps for connecting everything:

1. Connect the sensor's voltage (VDD) pin to Arduino Uno's 3.3 volts pin. Do this if you are using an MLX90614ESF-DCA-000 or an MLX90614ESF-BAA-000 sensor. If you are using an MLX90614ESF-AAA-000 or an MLX90614KSF-ACC-000 sensor, connect it to Arduino Uno's 5 volts pin.

2. Connect the sensor's ground (VSS) pin to Arduino Uno's ground (GND) pin.

3. Connect the sensor's SDA pin to Arduino Uno's A4 analog port.

4. Connect the sensor's SCL pin to Arduino Uno's A5 analog port.

5. Connect one leg of the 0.1 microfarad capacitor to the sensor's voltage pin and the other capacitor's leg to the sensor's ground pin.

Once you connected the wires and components, upload and run the code to the Arduino Uno board called `peripheral.ino`, explained in the *Programming the I2C interface* section. It should be showing the temperature data on the Arduino IDE's Serial Monitor. Follow these steps for opening the Serial Monitor for the Arduino Uno:

1. Open **Tools** from the IDE's main menu.

2. Select the **Serial Monitor** option.

3. From the Serial Monitor, make sure to select **9600** bauds.

4. Don't forget to select **Arduino Uno** from **Tools | Board**.

5. Select the right USB port where the Arduino Uno is connected, clicking on **Tool | Port** from the Arduino IDE.

The next section describes how to connect a Blue Pill to the Arduino Uno using the I2C bus and how to transmit temperature data from the Arduino Uno to the Blue Pill.

Connecting the Blue Pill to the Arduino Uno

This section shows how to connect the Arduino Uno and the Blue Pill through the I2C bus. The Arduino Uno will send the IR temperature data to the Blue Pill. Remember that the Blue Pill works as a controller (master) and the Arduino Uno is the peripheral (slave). *Figure 12.7* shows a Fritzing diagram with the two microcontroller boards:

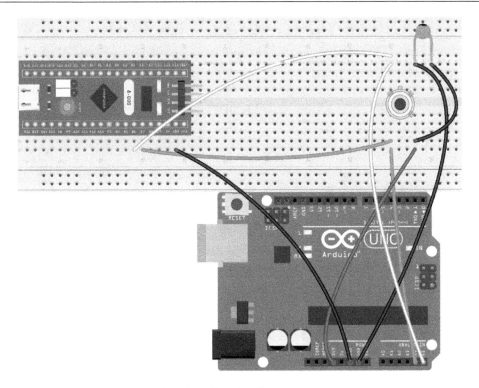

Figure 12.7 – The Blue Pill and Arduino Uno I2C connections

As *Figure 12.7* shows, here are the steps for connecting the Blue Pill to the I2C bus:

1. Connect Arduino Uno's ground (GND) to Blue Pill's ground (G or GND).

2. Connect Blue Pill's B7 pin to MLX90614's SDA pin.

3. Connect Blue Pill's B6 pin to MLX90614's SCL pin.

As you can see from all the connections from *Figure 12.7*, the Blue Pill, the IR sensor, and the Arduino Uno board are all connected by the SDA and the SCL pins. This is the I2C bus in our application.

> **Important note**
> Make sure to connect Blue Pill's ground (G or GND) to Arduino Uno's ground (GND). This will allow for correct data transmission across the I2C bus between the two microcontroller boards.

The following section describes how to show the IR temperature measurements on an LCD using the I2C bus.

Showing the temperature on an LCD

This section describes how to display the IR temperature measurement on an LCD through the I2C bus. The temperature data is sent by the Arduino Uno to the Blue Pill, as shown in the previous section. *Figure 12.8* shows a Fritzing diagram containing the microcontroller boards, the LCD, and the IR temperature sensor:

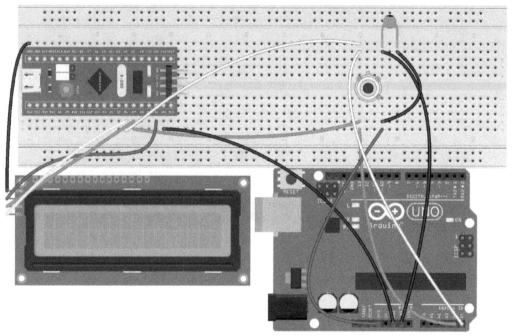

Figure 12.8 – The LCD connected to the I2C bus

As *Figure 12.8* shows, the LCD connection is simple. It requires four wires only, because the LCD used in this chapter is I2C-capable, having an I2C interface in the back. The following are the steps for connecting the LCD to the Blue Pill:

1. Connect the LCD's ground (GND) pin to Blue Pill's ground (G or GND) pin.

2. Connect the LCD's voltage (VCC) pin to Blue Pill's 5-volt (5V) pin.

3. Connect the LCD's SDA pin to Blue Pill's B7 pin.

4. Connect the LCD's SCL pin to Blue Pill's B6 pin.

Figure 12.9 depicts the back of the LCD, showing its I2C interface backpack attached to it:

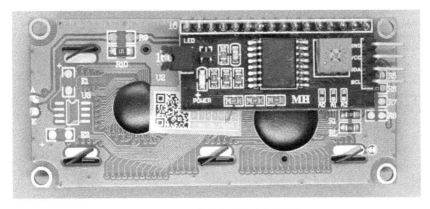

Figure 12.9 – The back of the LCD

From *Figure 12.9*, you can see that the LCD's I2C interface has a small variable resistor, just right of the big IC. You can rotate it to adjust the LCD contrast.

The LCD can display up to 16 characters on each of its two rows. This is enough to show the IR sensor temperature with two-digit precision and two decimals.

> **Important note**
> Make sure to connect the LCD's voltage (VCC) to Blue Pill's 5V. If you connect it to a 3.3-volt pin, it may not work properly.

Figure 12.10 shows how everything is connected:

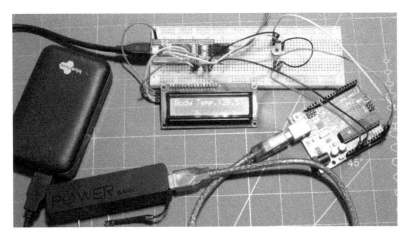

Figure 12.10 – The microcontroller boards, the sensor, and the LCD

Figure 12.10 shows that the LCD is showing a temperature of 28.53 degrees Celsius, because the IR sensor was unintentionally measuring the temperature of an LED lamp when the photo was taken! However, this circuit is intended to be used for body temperature measurement. The next section shows how to test the sensor by checking the temperature on different parts of the human body. We used two power banks connected to the Blue Pill and the Arduino Uno to try them out. If you finished the connection of the temperature sensor and the microcontroller boards, and if your LCD is displaying a temperature value, congratulations! You know how to use a touchless IR temperature sensor.

Testing the thermometer

In this section, we will test out how the IR sensor works as a thermometer by measuring the temperature of a human body. It seems that different body parts will get you slightly different temperature measurements. You should do a number of tests by measuring the body temperature from different parts, such as the forehead and the earlobe of a person. Remember that the distance between the sensor and the skin should be between 2 and 5 centimeters, although you should try out different distances and see what happens.

Before testing the thermometer, make sure that the skin is dry, clean, and unobstructed. In addition, confirm that the person has not been exposed to high heat, such as being out on a hot and sunny day, because this will change your measurements. If you are measuring skin temperature with the thermometer, make sure that the person is out of direct sunlight or you will get incorrect readings.

Medical studies indicate that the average skin surface temperature of the human body is in the range of 36.3 to 36.6 degrees Celsius. This is considered normal. However, a temperature higher than 37 degrees Celsius is suggestive of fever.

With these tips in mind, let's see how our thermometer works. *Figure 12.11* shows a person testing the IR sensor:

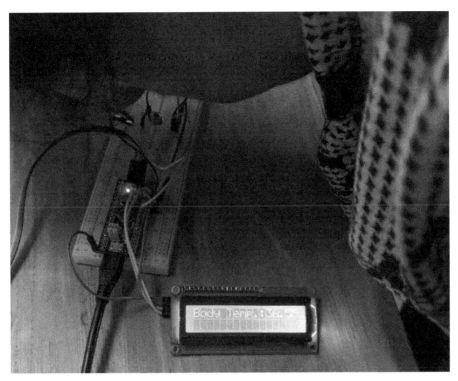

Figure 12.11 – Measuring the temperature

As *Figure 12.11* shows, the measured temperature was 36.55 degrees Celsius, which is within the normal range for adults. The tester needed to place her forefront close to the IR sensor to get reliable measurements. The distance between the forehead and the sensor can be increased by using a mirror tube placed on top of the sensor to steer the IR light through it.

Now that we have tested the chapter and realized that the thermometer works perfectly, let's recap in the *Summary* what we have learned.

Summary

In this chapter, you learned the basics of connecting an IR temperature sensor to a microcontroller board using the I2C serial data transmission protocol. As you could see, the I2C bus is an important part of the IR thermometer that we built in this chapter. This IR thermometer can be effective for checking human body temperature, like a regular thermometer will do. Since the temperature measurement is contactless, this can prevent the human body from touching the sensor and thus avoiding spreading viruses such as SARS-CoV-2.

This touchless thermometer may help to check whether a person has a fever and thus determine (along with other measurements) whether the person has contracted COVID-19. However, the IR temperature measurements explained in this chapter should not be definitive data to determine whether a person has COVID-19.

The next chapter explains another COVID-19-related project about measuring the recommended 2-meter distance between two people using an ultrasonic sensor.

Further reading

- Body Temperature (2020). Body temperature: What is (and isn't) normal? Available from `https://health.clevelandclinic.org/body-temperature-what-is-and-isnt-normal/`

- Gay, W. (2018) I2C, *In: Beginning STM32*, Apress, Berkeley, CA

- I2C (2021), *I2C tutorial.* Available from `https://learn.sparkfun.com/tutorials/i2c/all`

- Mankar, J., Darode, C., Trivedi, K., Kanoje, M., and Shahare, P. (2014), *Review of I2C protocol*, International Journal of Research in Advent Technology, 2(1)

13
COVID-19 Social-Distancing Alert

When the world celebrated the arrival of the year 2020, a pandemic was arising due to a new disease: COVID-19. With the emergence of this pandemic, all human activities were affected to a lesser or greater degree.

The education sector has been one of the most affected in this sense. All schools worldwide temporarily suspended their activities, since the risk of contagion in these environments can be very high. After a few months of lockdowns, schools around the world gradually began to resume face-to-face activities, following rigorous standards of disinfection and protocols to ensure physical distancing between students and school staff (Global Education Cluster, 2020).

The recommendation of the **World Health Organization** (**WHO**) for physical distancing is to remain at least 1 **meter** (**m**) (3 **feet** (**ft**)) apart between people, with 2 m (6 ft) being the most general recommendation to minimize the risk of contagion in children (KidsHealth, 2021). This measure is known as **social distancing**. Furthermore, technology sometimes acts as a great way to enforce these measures.

In this chapter, you will learn how to create a device that uses microcontroller technology to enforce social distancing to help children get used to maintaining a safe physical distance. When they are not at a safe physical distance, they will receive a sound alert from the device. The device you will create can be used by children as a wearable device for daily use by putting it in a case and using it as a necklace, as shown in the following screenshot:

Figure 13.1 – Wearable social-distancing device for children

In this chapter, we will cover the following main topics:

- Programming a piezoelectric buzzer
- Connecting an ultrasonic sensor to the microcontroller board
- Writing a program for getting data from the ultrasonic sensor
- Testing the distance meter

By completing this chapter, you will know how to program an electronic measurement of distance ranges using an **STM32 Blue Pill board**. You will also learn how to play an alarm when the distance is measured as less than 2 m.

> **Important note**
>
> This project is only for demonstration and learning purposes. Please do not use it as a primary social-distancing alarm for preventing the risk of COVID-19 contagion.

Technical requirements

The hardware components that will be needed to develop the social-distancing alarm are listed here:

- One solderless breadboard.
- One Blue Pill microcontroller board.
- One ST-LINK/V2 electronic interface is needed for uploading the compiled code to the Blue Pill board. Bear in mind that the ST-LINK/V2 interface requires four female-to-female jumper wires.
- One HC-SR04 ultrasonic sensor.
- One buzzer.
- Male-to-male jumper wires.
- Female-to-male jumper wires.
- A power source.
- Cardboard for the case.

As usual, you will require the Arduino **integrated development environment** (**IDE**) and the GitHub repository for this chapter, which can be found at `https://github.com/PacktPublishing/DIY-Microcontroller-Projects-for-Hobbyists/tree/master/Chapter13`

The Code in Action video for this chapter can be found here: `https://bit.ly/3gS2FKJ`

Let's begin!

Programming a piezoelectric buzzer

In this section, you will learn what a buzzer is, how to interface it with the **STM32 Blue Pill**, and how to write a program to build an alert sound.

A **piezoelectric buzzer** is a device that generates tones and beeps. It uses a piezoelectric effect, which consists of piezoelectric materials converting mechanical stress into electricity and electricity into mechanical vibrations. Piezoelectric buzzers contain a crystal with these characteristics, which changes shape when voltage is applied to it.

As has been common in these chapters, you can find a generic breakout module that is pretty straightforward to use, as shown in the following screenshot:

Figure 13.2 – Piezoelectric buzzer breakout board

This breakout board connects to the STM32 Blue Pill microcontroller board with three pins, outlined as follows:

- **Input/Output (I/O)**: This pin must be connected to a digital output of the microcontroller.

- **Voltage Common Collector (VCC)**: Pin to supply power to the sensor (**5 volts**, or **5V**).

- **Ground (GND)**: Ground connection.

Next, you will learn how to interface these pins with the Blue Pill microcontroller board.

Connecting the components

You will need a solderless breadboard to connect the buzzer to the STM32 Blue Pill microcontroller and a wire to connect the components. Follow these steps:

1. You need to place the STM32 Blue Pill and the buzzer into the solderless breadboard and leave space in the solderless breadboard to connect the jumper wires.

2. Connect the GND pin of the sensor to a GND terminal of the SMT32 Blue Pill.

3. Next, you need to connect the VCC pin to the 5V bus of the STM32 Blue Pill.

4. Finally, connect the I/O pin of the buzzer to pin B12 of the Blue Pill. The following screenshot shows all the components connected to the solderless breadboard:

Figure 13.3 – Piezoelectric buzzer interface to the Blue Pill

The following screenshot represents all the wiring between the STM32 Blue Pill and the piezoelectric buzzer and compiles the steps we just went through:

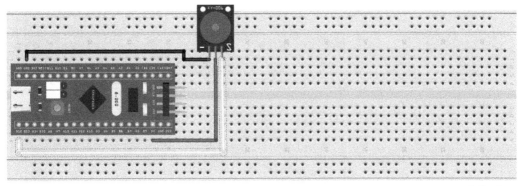

Figure 13.4 – Circuit for piezoelectric buzzer connection

Up to now, we have explored piezoelectric buzzers and their components and functionality. You have also learned how to connect them to an STM32 Blue Pill microcontroller board using a solderless breadboard.

Now, you are ready to write a program in the C language to reproduce an audible alert in the buzzer. Don't forget to use the STLink to upload the script to the STM32 Blue Pill microcontroller board.

Let's start developing a program to play an audible alert with the STM32 Blue Pill, as follows:

1. Let's get started defining which pin of the STM32 Blue Pill card pins will be used to play a sound in the buzzer. Run the following code:

    ```
    const int pinBuzzer = PB12;
    ```

    The selected pin was the PB12 pin (labeled B12 on the Blue Pill).

2. Next, we will leave the setup() part empty. You will not need to initialize code for this script.

3. The complete code is in the loop() part, as illustrated in the following code snippet:

    ```
    void loop()
    {
      tone(pinBuzzer, 1200);
      delay(250);
      noTone(pinBuzzer);
      delay(500);
      tone(pinBuzzer, 800);
      delay(250);
      noTone(pinBuzzer);
      delay(500);
    }
    ```

We are using two new functions: tone() and noTone(). Let's see what their functionality is.

tone() generates a square wave with a specific frequency from a pin. Its syntax is tone(pin, frequency, duration), where the pin parameter is the pin of the Blue Pill to which the buzzer is connected. frequency is the frequency of the tone in **Hertz (Hz)** of type unsigned int. The duration parameter is the tone's duration in **milliseconds (ms)**; it is an optional value and is of the unsigned long type.

noTone() stops the generation of the square wave that was started with tone(). An error will not be generated if a tone has not been previously generated. Its syntax is noTone(pin), where pin is the pin that is generating the tone.

So, the preceding code starts a 1,200 Hz tone and holds it for 250 ms with the delay() function. Later, it stops it and waits 500 ms to generate a new tone during 250 ms, now 800 Hz, and stops it again with the same 500-ms pause. These steps are repeated as long as the program is running to simulate an alert sound.

The code for this functionality is now complete. You can find the complete sketch in the Chapter13/buzzer folder in the GitHub repository.

Let's view how we have advanced our learning. We discovered a component to play tones, learned how to connect it to the STM32 Blue Pill microcontroller, and wrote the code to play an audible alert.

The skills you have acquired so far in this section will allow you to create other electronic systems that require play and audible alerts. Coming up next, we will learn about ultrasonic sensors.

Connecting an ultrasonic sensor to the microcontroller board

Before moving ahead, we need to learn about the functionality of the HC-SR04 ultrasonic sensor, how to interface it with the **STM32 Blue Pill**, and how to write a program to measure the distance between the sensor and another object.

This sensor emits an ultrasonic wave. When this wave collides with an object, the wave is reflected and received by the sensor. When the reflected signal is received, the sensor can calculate the time it took to be reflected, and thus the distance of the collision object can be measured.

The sensor can be seen in the following screenshot:

Figure 13.5 – Ultrasonic sensor

This sensor board connects to the STM32 Blue Pill microcontroller board with four pins, outlined as follows:

- **Trigger**: This pin enables the ultrasonic wave.

- **Echo**: This pin receives the reflected wave.

- **VCC**: The pin to supply power to the sensor (5V).

- **GND**: Ground connection.

Next, it's time to interface these pins with the Blue Pill microcontroller.

Connecting the components

A solderless breadboard will be required to connect the buzzer to the STM32 Blue Pill microcontroller and wire to connect the components. Proceed as follows:

1. You need to place the STM32 Blue Pill and the sensor into the solderless breadboard and leave space to connect the jumper wires.

2. Connect the GND pin of the sensor to a GND terminal of the SMT32 Blue Pill.

3. Next, you need to connect the VCC pin to the 5V bus of the STM32 Blue Pill.

4. Finally, connect the trigger pin of the buzzer to pin C14 and the echo pin to the C13 pin of the Blue Pill. The following screenshot shows all the components connected to the solderless breadboard:

Figure 13.6 – Piezoelectric buzzer interface to the Blue Pill

The following screenshot represents all the wiring between the STM32 Blue Pill and the ultrasonic sensor:

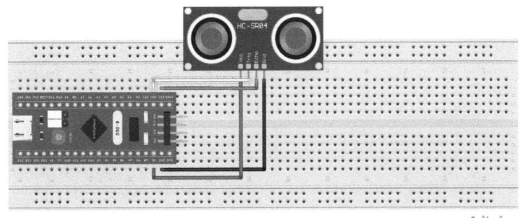

Figure 13.7 – Circuit for the ultrasonic sensor connection

Up to now, you have learned how to connect a sensor to the STM32 Blue Pill microcontroller board using a solderless breadboard.

Now, you will learn how to write a program in the C language to reproduce an audible alert in the buzzer. Don't forget to use the STLink to upload the script to the STM32 Blue Pill microcontroller board.

Writing a program for getting data from the ultrasonic sensor

In this section, you will learn how to write a program to gather data from the ultrasonic sensor. Let's start, as follows:

1. First, we will define which pins of the STM32 Blue Pill card will be used to read the sensor data. Also, we will declare two variables to save the duration of the sound-wave travel and another for calculating the distance traveled, as illustrated in the following code snippet:

```
const int pinTrigger = PC14;
const int pinEcho = PC13;
long soundWaveTime;
long distanceMeasurement;
```

The selected pins were the PC13 and PC14 pins (labeled C13 and C14 on the Blue Pill).

2. Next, in the setup() function, begin the serial communication. You will set the trigger pin as an output pin and the echo pin as an input pin. We need to initialize the trigger in the LOW value. The code is illustrated in the following snippet:

```
void setup() {
  Serial.begin(9600);
  pinMode(pinTrigger, OUTPUT);
  pinMode(pinEcho, INPUT);
  digitalWrite(pinTrigger, LOW);
}
```

3. Now, we will code the loop() function. We need to start the ultrasonic wave, wait 10 **seconds (sec)**, and turn off the wave. The code is illustrated in the following snippet:

```
void loop()
{
  digitalWrite(pinTrigger, HIGH);
  delayMicroseconds(10);
  digitalWrite(pinTrigger, LOW);
  ...
}
```

4. The next step is to read the echo pin of the sensor to know the total travel time of the wave. We do this with the `pulseIn()` function and store it in the variable we declared at the beginning, for this purpose. To calculate the distance, we take the value of the return pulse and divide it by 59 to obtain the distance in **centimeters** (**cm**), as illustrated in the following code snippet:

```
void loop()
{
   digitalWrite(pinTrigger, HIGH);
   delayMicroseconds(10);
   digitalWrite(pinTrigger, LOW);
   soundWaveTime = pulseIn(pinEcho, HIGH);
   distanceMeasurement = soundWaveTime/59;
   ...
}
```

5. Finally, you will show the distance value between the sensor and any object in front of our device in the serial console, as follows:

```
void loop()
{
   digitalWrite(pinTrigger, HIGH);
   delayMicroseconds(10);
   digitalWrite(pinTrigger, LOW);
   soundWaveTime = pulseIn(pinEcho, HIGH);
   distanceMeasurement = soundWaveTime/59;
   Serial.print("Distance: ");
   Serial.print(distanceMeasurement);
   Serial.println("cm");
   delay(500);
}
```

The code for this functionality is now complete. You can find the complete sketch in the `Chapter13/ultrasonic` folder in the GitHub repository.

At the end of this section, you have learned how to write a program in the C language to measure the distance between an object and an ultrasonic sensor connected to the STM32.

With these skills, you will be able to develop electronic projects that require distance measurement, such as car-reverse-impact prevention.

Testing the distance meter

Before testing the distance meter, we will need to wire together the buzzer and the ultrasonic sensor to the SMT32 Blue Pill in the solderless breadboard. The following screenshot illustrates a complete circuit diagram including the STM32, ultrasonic sensor, and buzzer together in the solderless breadboard:

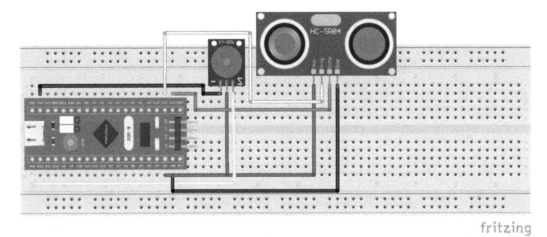

Figure 13.8 – Full circuit diagram of our social-distancing device

The following screenshot shows how everything should be connected in the actual system:

Figure 13.9 – The buzzer and ultrasonic sensor connections

Now, to complete the connection of the complete social-distancing device, we will need to write a new script combining the Chapter13/buzzer and Chapter13/ultrasonic scripts. The new script will be named Chapter13/distance_meter. Follow these steps:

1. We need to declare the constants and variables of both scripts and add a new script to define the safety distance between the sensor device and another object. The code to do this is illustrated in the following snippet:

```
const int pinTrigger = PC14;
const int pinEcho = PC13;
const int pinBuzzer = PB12;
const int distanceSafety = 200;
long soundWaveTime;
long distanceMeasurement;
```

For COVID-19 social distancing, we will use 200 cm (2 m).

2. The setup() function remains the same as the ultrasonic script, as illustrated in the following code snippet:

```
void setup() {
  Serial.begin(9600);
  pinMode(pinTrigger, OUTPUT);
  pinMode(pinEcho, INPUT);
  digitalWrite(pinTrigger, LOW);
}
```

3. Finally, in the loop() function, we will include a conditional to verify if our social-distancing device is physically separated less than 2 m from another person (object). If this is the case, play the audible alert. Here is the code to do this:

```
void loop()
{
  digitalWrite(pinTrigger, HIGH);
  delayMicroseconds(10);
  digitalWrite(pinTrigger, LOW);
  soundWaveTime = pulseIn(pinEcho, HIGH);
  distanceMeasurement = soundWaveTime/59;
  Serial.print("Distance: ");
  Serial.print(distanceMeasurement);
  Serial.println("cm");
```

```
delay(500);
if (distanceMeasurement < distanceSafety) {
  Serial.println("Sound alert");
  tone(pinBuzzer, 1200);
  delay(250);
  noTone(pinBuzzer);
  delay(500);
  tone(pinBuzzer, 800);
  delay(250);
  noTone(pinBuzzer);
  delay(500);
}
}
```

Now, you can measure social distancing, and it can be possible to use our device as a necklace in schools to maintain a safe physical distance, *only as a complement to the official safety instructions.*

To achieve this, we can create a cardboard case and insert our device in it. Print the template shown in the following screenshot—you can download this from the Chapter13/cardboard GitHub folder:

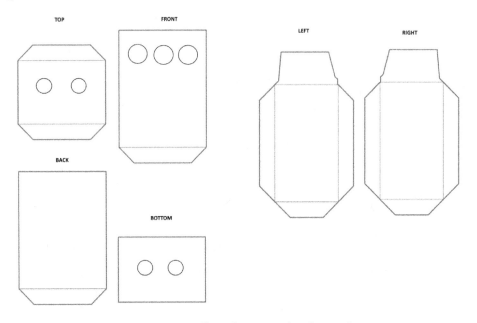

Figure 13.10 – A cardboard-case template for our device

To better fit our electronic device in the case, it is recommended to change the jumper wires used to build the prototype (male-to-male) to male-to-female jumper wires and power it with a 5V battery, as shown in the following screenshot:

Figure 13.11 – Adapted connections to fit into a cardboard-case template

Finally, cut and glue the case and put the device we just created into the case to create a wearable device, as shown in the following screenshot:

Figure 13.12 – A cardboard-case template for our device

Using this device, you will know whether you are at a safe distance to avoid possible COVID-19 infections by droplets.

We have reached the end of *Chapter 13*. Congratulations!

Summary

What did we learn in this project? Firstly, we learned how to connect a piezoelectric buzzer to our Blue Pill microcontroller board and code a program to play an audible alarm. Then, we wrote a program to measure the distance between our electronic device and another object.

We also learned how to combine the two projects to create a social-distancing device that can be used to maintain a safe physical distance in this COVID-19 pandemic—for example, in schools, because children are more distracted and are more friendly and sociable.

It is important to remind you that this project is intended for learning purposes and should not be used as a primary alarm for preventing the risk of COVID-19 contagion in any circumstances. This is mainly because, at this time, we know that the main risk is airborne.

In the next chapter, we will learn to build a 20-second hand-washing timer.

Further reading

- Global Education Cluster. *Safe back to school: a practitioner's guide.* UNESCO. 2020:

  ```
  https://healtheducationresources.unesco.org/library/
  documents/safe-back-school-practitioners-guide
  ```

- KidsHealth. *Coronavirus (COVID-19): Social Distancing With Children.* 2021:

  ```
  https://kidshealth.org/en/parents/coronavirus-social-
  distancing.html
  ```

14
COVID-19 20-Second Hand Washing Timer

This chapter describes a useful project where you will make a touchless timer by waving at an ultrasonic sensor. This timer will count the minimum time of 20 seconds recommended by health authorities for properly washing our hands for preventing contamination from viruses such as SARS-CoV-2 that produces COVID-19 symptoms. The project involves an inexpensive ultrasonic sensor that detects when a user waves at the sensor by measuring the distance between the user and the circuit, triggering the counting. This application must be enclosed in a waterproof container to avoid soaking the circuit while the user washes their hands and damaging it. We explain at the end of the chapter how to do this.

In this chapter, we will cover the following main topics:

- Programming the counter (timer)
- Showing the timer on an LCD
- Connecting an ultrasonic sensor to the microcontroller board
- Putting everything together – think of a protective case for the project!
- Testing the timer

By the end of this chapter, you will have learned how to properly connect an ultrasonic sensor and an LCD to a microcontroller board. In addition, you will learn how to read input values from a sensor to activate the 20-second counting. You will also learn how to code an efficient and effective timer that runs on a microcontroller board.

Technical requirements

The software tool that you will be using in this chapter is the Arduino IDE for editing and uploading your programs to the Blue Pill microcontroller board.

The code used in this chapter can be found in the book's GitHub repository:

`https://github.com/PacktPublishing/DIY-Microcontroller-Projects-for-Hobbyists/tree/master/Chapter14`

The Code in Action video for this chapter can be found here: `https://bit.ly/3gQZdPf`

In this chapter, we will use the following pieces of hardware:

- One solderless breadboard.
- One Blue Pill microcontroller board.
- One micro-USB cable for connecting your microcontroller board to a computer and a power bank.
- One USB power bank.
- One ST-INK/V2 electronic interface, needed for uploading the compiled code to the Blue Pill. Bear in mind that the ST-LINK/V2 requires four female-to-female DuPont wires.
- One HC-SR04 ultrasonic sensor.
- One 1602 16x2 LCD.
- One 2.2k ohm resistor, 1/4 watt. This is for the LCD.
- A dozen male-to-male and a dozen male-to-female DuPont wires.

The next section explains how to code the 20-second timer that runs on the Blue Pill microcontroller board.

Programming the counter (timer)

This section shows you how to code the 20-second timer. Programming a timer like this one is not trivial, since a user could activate the timer many times by waving at the ultrasonic sensor more than once while the counter is on. The program should not take into account those multiple activations if the 20-second counting is going on; otherwise, the counter will re-start multiple times and the counting will not be accurate. We also need to consider saving energy by turning off the LCD when not in use.

We should code our 20-second timer application by following these steps:

1. Read values from the ultrasonic sensor.

2. Check whether the user is waving at the sensor within 15 centimeters of the sensor. If this happens, do this:

    a) Turn on the LCD light.

    b) Show the **Lather hands** message and show the 20-second counting on the LCD.

    c) When the counter finishes, show this message on the LCD: **Rinse your hands**.

    d) Wait for 4 seconds and then turn off the LCD to conserve energy.

    e) Return to *step 1*.

The following is the Arduino IDE code that runs on the Blue Pill, programmed following the preceding steps. The following code snippet shows how the variables and constants are defined. The HC-SR04's echo digital value is obtained from port A9 (labeled PA9) and stored in the echo variable. Similarly, the trigger value is sent to the ultrasonic sensor through port A8 (PA8) and stored in the trigger variable. Please note that the dist_read variable stores the distance between an object (for example, a waving hand) and the ultrasonic sensor, measured in centimeters:

```
#include "elapsedMillis.h"
elapsedMillis timeElapsed;
#include <LiquidCrystal.h>
const int rs = PB11, en = PB10, d4 = PB0, d5 = PA7, d6 =
    PA6, d7 = PA5;
LiquidCrystal lcd(rs, en, d4, d5, d6, d7);
#define backLight PB12
#define trigger PA8
#define echo PA9
const int max_dist=200;
float duration,dist_read=0.0;
```

As you can see from the preceding code, it uses a library called `elapsedMillis.h` to calculate the time in milliseconds that has elapsed on the counting. This library can also be obtained from `https://www.arduino.cc/reference/en/libraries/elapsedmillis/`. You can find this library in the code folder uploaded to the book's GitHub repository. It is useful because by using this library we avoid using the `delay()` function in the 20-second counting. If we use the `delay()` function, the Blue Pill's counting and reading sensor values from the ports could mess up. Note that the library is written between double quotes because in C++ this means that the library is in the same folder as the source code, which is a library that does not belong to the original Arduino IDE installation. The code also uses the `LiquidCrystal.h` library, used for controlling the 1602 LCD. This library already comes with the standard Arduino IDE installation so there is no need for installing it separately.

This code snippet sets up the LCD and the Blue Pill ports:

```
void setup() {
  lcd.begin(16, 2);
  pinMode(trigger, OUTPUT);
  pinMode(echo, INPUT);
  pinMode(backLight, OUTPUT);
}
```

The following code segment shows the code's main loop, which reads the ultrasonic sensor values and calculates the distance between the user's waving hand and the sensor:

```
void loop() {
  digitalWrite(trigger, LOW);
  delayMicroseconds(2);
  digitalWrite(trigger, HIGH);
  delayMicroseconds(10);
  digitalWrite(trigger, LOW);
  duration = pulseIn(echo, HIGH);
  dist_read = (duration*.0343)/2;
```

The following code segment from the main loop function calculates whether the distance between the user and the sensor is equal to or less than 15 centimeters, then activates the 20-second counter and shows it on the LCD:

```
  if ((dist_read<=15) & (dist_read>0))
  {
    lcd.display();
```

```
    digitalWrite(backLight, HIGH);
    timeElapsed=0;
    lcd.setCursor(0, 0);
    lcd.print("lather hands :) ");
    lcd.setCursor(0, 1);
    lcd.print("  ");
    while (timeElapsed < 21000)
    {
        lcd.setCursor(0, 1);
        lcd.print(timeElapsed / 1000);
    }
    lcd.setCursor(0, 0);
    lcd.print("rinse hands :)   ");
    delay(4000);
  }
  lcd.noDisplay();
  digitalWrite(backLight, LOW);
}
```

> **Tip**
> You can also run the previous code on Arduino microcontroller boards.
> You just need to change the port numbers used for the LCD and sensor
> connections. For example, if you are using an Arduino Uno board, change this
> line to `const int rs=12,en=11,d4=5,d5=4,d6=3,d7=2;` using
> Arduino board digital ports 12, 11, 5, 4, 3, and 2, respectively. You will also
> need to change these lines to the following:
>
> `#define backLight 6`
>
> `#define trigger 7`
>
> `#define echo 8`
>
> So, you will use Arduino digital ports 6, 7 and 8 for the ultrasonic sensor.

Bear in mind that the code uploaded to the GitHub repository contains many comments explaining its most important parts.

The next section explains how to connect the 1602 LCD to the Blue Pill to show the 20-second count on it.

Showing the timer on an LCD

In this section, we explain how to connect and use the 1602 LCD to show the timer on it. *Figure 14.1* shows the Fritzing diagram similar to the one explained in *Chapter 5, Humidity and Temperature Measurement*:

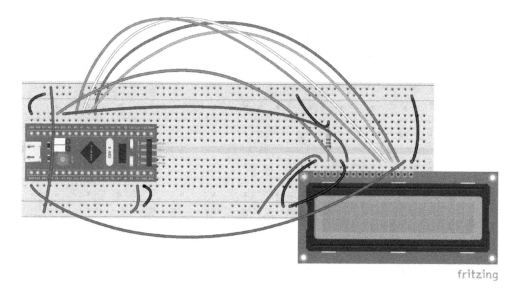

Figure 14.1 – The LCD connected to the Blue Pill microcontroller board

The following are the steps for connecting the LCD to the Blue Pill, following the diagram from *Figure 14.1*:

1. Connect the Blue Pill's **GND** (also labeled as **G**) pins to the solderless breadboard rails.

2. Connect the Blue Pill's **5V** pin (providing 5 volts) to the breadboard rails.

3. Connect the USB cable to the Blue Pill and then to your computer or a USB power bank.

4. Insert the LCD's 16 pins into the solderless breadboard.

5. Connect the LCD's **VSS** pin to ground (the lower breadboard rail).

6. Connect the LCD's **VDD** pin to 5 volts (the lower breadboard rail).

7. Connect the 2.2k ohm resistor to the LCD's **V0** pin and to ground (the lower breadboard rail).

8. Connect the LCD's **RS** pin to the Blue Pill's **B11** pin.

9. Connect the LCD's **RW** pin to ground (lower breadboard rail).

10. Connect the LCD's **E** pin to the Blue Pill's **B10** pin.

11. Connect the LCD's **D4** pin to the Blue Pill's **B0** pin.

12. Connect the LCD's **D5** pin to the Blue Pill's **A7** pin.

13. Connect the LCD's **D6** pin to the Blue Pill's **A6** pin.

14. Connect the LCD's **D7** pin to the Blue Pill's **A5** pin.

15. Connect the LCD's **A** pin to the Blue Pill's port **B12**.

16. Connect the LCD's **K** pin to ground (lower breadboard rail).

17. The LCD's **D0**, **D1**, **D2**, and **D3** pins are not connected.

After doing the preceding steps, you have accomplished connecting the LCD to the Blue Pill board. The LCD will be useful for showing the 20-second count. Well done!

Figure 14.2 shows how everything is connected:

Figure 14.2 – The Blue Pill microcontroller board connected to the LCD

As you can see from *Figure 14.2*, the 1602A LCD is easy to connect to the Blue Pill. The 2.2k ohm connected to the LCD's pin **V0** sets up the LCD's contrast.

> **Tip**
> You can use a 50k ohm variable resistor (also known as a **potentiometer**) instead of the 2.2k ohm resistor connected to the LCD's pin **V0** to adjust the display contrast. Just connect the potentiometer's middle pin to **V0**, one pin to ground, and the other pin to **5V**.

Please note that the LCD pin **A** (pin no. 15) is connected to the Blue Pill's **B12** port, which controls the LCD by turning its back light on or off via coding.

> **Tip**
> Allow enough space on the solderless breadboard between the Blue Pill and the rest of the electronic components (LCD, ultrasonic sensor, and so on) to facilitate the connection of the ST-Link/V2 interface to the Blue Pill.

The next section explains how to use the ultrasonic sensor to see whether a user is waving at the sensor to trigger the 20-second timer.

Connecting an ultrasonic sensor to the microcontroller board

This section explains how an ultrasonic sensor works, and it describes how to connect the HC-SR04 sensor to the Blue Pill microcontroller board, describing how to use its four-pin functions. The ultrasonic sensor will be used to check whether the user waves at it to initiate the 20-second counting.

What is an ultrasonic sensor?

Ultrasonic waves are sound waves that have a frequency that is higher than the frequencies that most human beings can hear, which is above 20,000 Hz. Ultrasonic sounds, or ultrasound, can have different applications, including something called **echolocation**, used by animals such as bats for identifying how far their prey is using reflected sounds. The same principle is applied in ultrasonic sensors.

An **ultrasonic sensor** is a dedicated electronic component that generally contains a number of electronic parts such as resistors, transistors, diodes, a crystal clock, a special microphone, and a speaker. Many ultrasonic sensors are technically modules, because they integrate a number of electronic parts, and this integration as a module facilitates the connection with other devices such as microcontroller boards. An ultrasonic sensor measures the distance between an object (for example, a waving hand) and the sensor by using ultrasonic sound waves. The sensor emits ultrasonic sound waves through a speaker and receives through a microphone the reflected ultrasonic waves that hit the object. The sensor measures the time it takes between the sound wave emission and reception.

How does an ultrasonic sensor work?

An ultrasonic sensor (such as the **HC-SR04**) emits and receives ultrasonic sound waves (working like **sonar**) to determine the distance to an object. Sonar is an echolocation device used for detecting objects underwater, emitting sound pulses (generally using ultrasound frequencies), measuring the time it takes for the reflection of those pulses, and calculating the distance between the object and the sonar device. The ultrasonic sensor used in this chapter is not meant to be used underwater, as we can see later.

Some ultrasonic sensors (such as the HC-SR04 used in this chapter) use sound waves with a frequency of 40 kHz, well above the range of sound frequencies that the human ear can perceive on average, which is 20 Hz to 20 kHz.

This is how the HC-SR04 ultrasonic sensor works:

1. A microcontroller board sends a digital signal to the sensor's `Trig` pin, triggering (initiating) the ultrasonic wave emission through the sensor's speaker.

2. When a high-frequency sound wave hits an object, it is reflected back to the sensor and this reflection is picked up by the sensor's microphone.

3. The sensor sends out a digital signal to the microcontroller board through its `Echo` pin.

4. A microcontroller board receives that digital signal from the `Echo` pin, encoding the duration between the sound wave emission and reception.

The distance between the sensor and the object is calculated as follows:

$$D=(T^*C)/2$$

The symbols denote the following:

- **D**: Distance
- **T**: Time it takes between ultrasonic wave emission and reception (duration)
- **C**: General speed of sound (343 m/s in dry air)

The distance is divided by 2 because we need just the sound wave's return distance.

Figure 14.3 shows a Fritzing diagram of the ultrasonic sensor:

Figure 14.3 – The HC-SR04 ultrasonic sensor pinout

From *Figure 14.3*, you can see the sensor pinout. The **VCC** pin is connected to a 5-volt power supply. The **Trig** and **Echo** pins are connected to the microcontroller board's digital ports. The **GND** pin is connected to ground.

Here are the technical characteristics of the HC-SR04 sensor:

- **Operating voltage**: DC 5 volts
- **Operating frequency**: 40 kHz
- **Operating current**: 15 mA
- **Maximum operational range**: 4 meters
- **Minimum operational range**: 2 centimeters
- **Resolution**: 0.3 centimeters
- **Measuring angle**: 30 degrees (sensor's field of view)

Figure 14.4 shows the HC-SR04 ultrasonic sensor used in this chapter:

Figure 14.4 – The HC-SR04 ultrasonic sensor

From *Figure 14.4*, you can see that the sensor has two speaker-like components. One of them is actually a small speaker emitting ultrasonic sound signals and the other one is a microphone that captures those signals back after they bounce on an object. *Figure 14.5* shows the back of the HC-SR04 ultrasonic sensor:

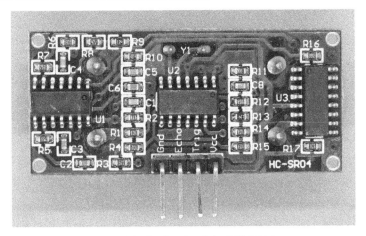

Figure 14.5 – The back side of the HC-SR04 sensor

As you can see from *Figure 14.5*, the back of the sensor contains electronic components such as resistors, transistors, and integrated circuits that support the generation and reception of ultrasonic signals.

There are other types of ultrasonic sensors, such as the Maxbotix MaxSonar ultrasonic sensor. Its Fritzing diagram is shown in *Figure 14.6*:

Figure 14.6 – The Maxbotix MaxSonar ultrasonic sensor

The sensor shown in *Figure 14.6* can be used with microcontroller boards. The Maxbotix MaxSonar is an accurate and long-range ultrasonic sensor (it can measure distances up to 6.45 meters), but it is expensive and requires connecting seven wires to its seven pins. The HC-SR04 sensor will suffice for our 20-second timer application. It is low cost and easy to connect to, requiring only four wires.

The timer starts when the ultrasonic sensor detects a user waving at it, so the sensor will trigger the timer.

Figure 14.7 shows how to connect the HC-SR04 ultrasonic sensor to the Blue Pill:

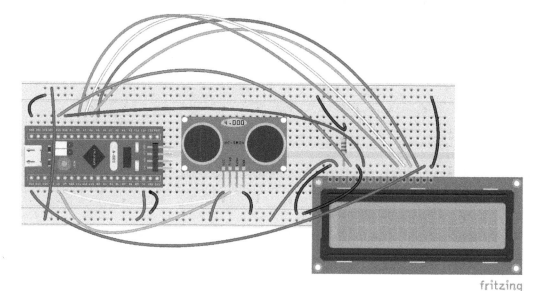

Figure 14.7 – The ultrasonic sensor connected to the Blue Pill microcontroller board

From *Figure 14.7*, you can see that one of the breadboard's lower rails is connected to the Blue Pill's **5V** pin. The other rail is connected to Blue Pill's ground. The following are the steps for connecting the HC-SR04 sensor to the Blue Pill in addition to the steps followed in *Figure 14.1*, following the diagram from *Figure 14.7*:

1. Connect the sensor's **VCC** pin to the breadboard's low rail that is connected to **5V**.

2. Connect the sensor's **Trig** pin to the Blue Pill's **A8** pin.

3. Connect the sensor's **Echo** pin to the Blue Pill's **A9** pin.

4. Connect the sensor's **GND** pin to the breadboard's lower rail that is connected to ground.

Figure 14.8 shows how everything is connected:

Figure 14.8 – The Blue Pill connected to the LCD and the ultrasonic sensor

Please note from *Figure 14.8* that all the connections from the ultrasonic sensor are done on its back side. This is to avoid any cable obstructing the ultrasonic signals sent and received on the front of the sensor. Also note that the 1602 LCD's power (pin **VDD**) is connected to Blue Pill's pin **5V**. If you feed the LCD with 3.3 volts, it may not work.

> **Tip**
> Make sure that there are no wires obstructing the field of view of HC-SR04 ultrasonic sensor; otherwise, they will produce erratic or false measurements and results with the 20-second counting.

Also note from *Figure 14.8* that the breadboard's upper rail is connected to the Blue Pill's **G** pin (in some Blue Pills it is labeled as **GND**), which serves to connect the LCD's ground and its 2.2k ohm resistor that is used to preset the LCD's contrast.

The next section explains how to encase the whole project to protect it from dust, water, and so on, and to facilitate its use in a place for washing hands.

Putting everything together – think of a protective case for the project!

This section shows how you can place the electronic circuit with the ultrasonic sensor inside a protective case. The section also shows some suggestions on how to fit everything in a plastic or glass container, because if you use the 20-second counter in a bathroom or in a place close to a hand washing sink, you will need to protect the circuit against water spilling and soap stains that can damage the electronic components used in this 20-second counter project. We do not recommend you connect the Blue Pill board to a wall USB adapter for security reasons. It is best to connect the Blue Pill to a USB power bank.

If you can't fit the whole 20-second counter circuit (including its solderless breadboard) in a plastic or glass container, try connecting the Blue Pill on a smaller solderless breadboard such as a half breadboard. Detach the ultrasonic sensor and the LCD from the breadboard and position and attach them inside the container with strong adhesive tape. A Fritzing diagram about this smaller circuit is shown in *Figure 14.9*:

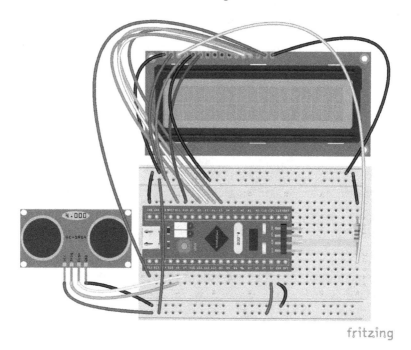

Figure 14.9 – Connecting everything on a small solderless breadboard

As you can see from *Figure 14.9*, by using a half-breadboard you can make all the connections more compact so you can use a small container such as an empty instant coffee Jar or any plastic container that has a lid. You could mount the ultrasonic sensor on the lid and place the rest inside of the container or jar. You will need to use female-to-male DuPont wires to connect the LCD and the ultrasonic sensor to the half breadboard.

Figure 14.10 shows a prototype design with a custom-made case giving you an idea of how you could encase all the components and protect the electronics from water spilling:

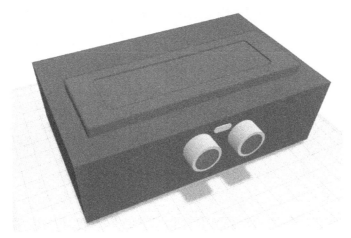

Figure 14.10 – A 3D prototype design containing the whole project

As you can see from *Figure 14.10*, that case could be a plastic box where the 1602A LCD is placed on top of it. The HC-SR04 sensor is placed at the front of the case. The interior of the case could contain the small breadboard, wires, the Blue Pill, the resistor, the power bank, and the USB cable. Don't place the whole project very close to a hand wash sink, just in case.

Testing the timer

This section shows how to test out the 20-second timer.

Once you insert the electronic circuit with the sensor, the Blue Pill, and the LCD in a protective case, try it in a bathroom. Carefully place it close to a hand washing sink if you can, to facilitate activating it and seeing the counting while you wash your hands. See whether you can fix it to a wall or a surface so it won't move and that no one accidentally knocks it over while waving at it. Safety first!

You should connect the Blue Pill to a portable power bank that has a USB socket. This is to avoid connecting the Blue Pill to a wall USB adapter to make it safer to use in an environment such as a bathroom, as shown in *Figure 14.11*:

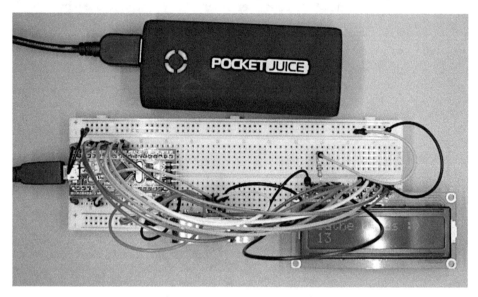

Figure 14.11 – A power bank connected to the Blue Pill microcontroller board

You can test out everything with a small power bank, such as the one shown in *Figure 14.11*.

Try activating the timer by waving at the sensor numerous times. You will see that sometimes the circuit counts up to 21. This is because most microcontroller boards (including the Blue Pill) do not calculate the time very accurately. Try adding a variable and a decision to the code to show the count up to number 20. Hint: Stop showing the counting when it reaches 20. It is not critical if it counts up to 21 when you wash your hands for trying to *destroy* the virus that causes COVID-19. The longer the counting the better.

You can adjust the detected distance if you feel that the user needs to wave at the sensor at a different distance. Try changing the value from this line of code:

```
if ((dist_read<=15) & (dist_read>0))
```

The 15 value means that the LCD will activate and show the counting if you are waving at the sensor at a distance of 15 centimeters or less from the sensor. Try to change the value to a greater number, perhaps 20 centimeters.

If you think that the text and numbers shown on the LCD need more contrast, try changing the 2.2k ohm resistor to a smaller one, such as 1 k ohm. This may happen if your bathroom or the place where you will use the counter is too bright.

This is an interesting test: try the 20-second counter with people of different ages, to see whether the ultrasonic sensor can detect different hand sizes. For example, see whether the sensor detects the waving hands of small children and adults.

Summary

In this chapter, we learned the basics of coding an easy-to-read 20-second counter. This count is recommended by many health authorities for properly washing our hands during that time in an attempt to destroy some viruses such as the one that causes COVID-19. The chapter also explained how the HC-SR04 ultrasonic sensor works for activating the counter. One major skill that you gained on completing the project from this chapter is that you learned how to connect a practical LCD to a microcontroller board, and how we could show the counting on an LCD. You can use the LCD in other projects that require showing numeric or text data from a microcontroller board.

We have covered in this chapter a practical way to obtain data from a sensor, process it on the microcontroller board, and do something about it such as showing results on an LCD. Obtaining data from sensors and processing it is one of the main applications of microcontrollers, leveraging their simplicity for connecting sensors to their input/output ports.

Further reading

- Choudhuri, K. B. R. (2017), *Learn Arduino Prototyping in 10 Days*, Birmingham, UK: Packt Publishing Ltd

- Gay, W. (2018), *Beginning STM32: Developing with FreeRTOS, libopencm3 and GCC*, New York, NY: Apress

- HC-SR04 (2013), *HC-SR04 user's manual V1.0. Cytron Technologies*, available from `https://docs.google.com/document/d/1Y-yZnNhMYy7rwhAgyL_pfa39RsB-x2qR4vP8saG73rE/edit?usp=sharing`

- Horowitz, P. and Hill, W. (2015), *The Art of Electronics*, [3rd ed.] Cambridge University Press: New York, NY

- LCD1602 (2009), *LCM module data sheet TC1602A-01T*, Tinsharp Industrial Co., Ltd. available from `https://cdn-shop.adafruit.com/datasheets/TC1602A-01T.pdf`

- Microchip (2019), *PIC16F15376 Curiosity Nano hardware user guide*, Microchip Technology, Inc. available from `http://ww1.microchip.com/downloads/en/DeviceDoc/50002900B.pdf`

Packt.com

Subscribe to our online digital library for full access to over 7,000 books and videos, as well as industry leading tools to help you plan your personal development and advance your career. For more information, please visit our website.

Why subscribe?

- Spend less time learning and more time coding with practical eBooks and Videos from over 4,000 industry professionals

- Improve your learning with Skill Plans built especially for you

- Get a free eBook or video every month

- Fully searchable for easy access to vital information

- Copy and paste, print, and bookmark content

Did you know that Packt offers eBook versions of every book published, with PDF and ePub files available? You can upgrade to the eBook version at packt.com and as a print book customer, you are entitled to a discount on the eBook copy. Get in touch with us at customercare@packtpub.com for more details.

At www.packt.com, you can also read a collection of free technical articles, sign up for a range of free newsletters, and receive exclusive discounts and offers on Packt books and eBooks.

Other Books You May Enjoy

If you enjoyed this book, you may be interested in these other books by Packt:

Hands-On RTOS with Microcontrollers

Brian Amos

ISBN: 978-1-83882-6-734

- Get up and running with the fundamentals of RTOS and apply them on STM32
- Enhance your programming skills to design and build real-world embedded systems
- Get to grips with advanced techniques for implementing embedded systems

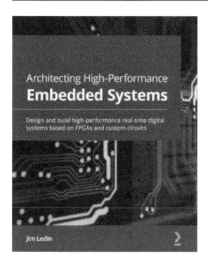

Architecting High-Performance Embedded Systems

Jim Ledin

ISBN: 978-1-78995-5-965

- Learn the basics of embedded systems and real-time operating systems
- Understand how FPGAs implement processing algorithms in hardware
- Design, construct, and debug custom digital systems from scratch using KiCad

Packt is searching for authors like you

If you're interested in becoming an author for Packt, please visit authors. packtpub.com and apply today. We have worked with thousands of developers and tech professionals, just like you, to help them share their insight with the global tech community. You can make a general application, apply for a specific hot topic that we are recruiting an author for, or submit your own idea.

Share Your Thoughts

Now you've finished *DIY Microcontroller Projects for Hobbyists*, we'd love to hear your thoughts! Scan the QR code below to go straight to the Amazon review page for this book and share your feedback or leave a review on the site that you purchased it from.

https://packt.link/r/1-800-56413-9

Your review is important to us and the tech community and will help us make sure we're delivering excellent quality content.

Index